INTRODUCTION TO
GLOBAL TECTONIC SYSTEMS

INTRODUCTION TO
GLOBAL TECTONIC SYSTEMS

Edited by

Yuzhu Kang

Sinopec Petroleum Exploration and Production Research Institute, China

World Scientific

NEW JERSEY · LONDON · SINGAPORE · BEIJING · SHANGHAI · HONG KONG · TAIPEI · CHENNAI · TOKYO

Published by

World Scientific Publishing Co. Pte. Ltd.

5 Toh Tuck Link, Singapore 596224

USA office: 27 Warren Street, Suite 401-402, Hackensack, NJ 07601

UK office: 57 Shelton Street, Covent Garden, London WC2H 9HE

Library of Congress Cataloging-in-Publication Data

Names: Kang, Yuzhu, editor.

Title: Introduction to global tectonic systems / edited by Yuzhu Kang,
    Sinopec Petroleum Exploration and Production Research Institute, China.

Description: New Jersey : World scientific, [2024] | Includes bibliographical references and index.

Identifiers: LCCN 2023047500 | ISBN 9789811285554 (hardcover) |
    ISBN 9789811285561 (ebook for institutions) | ISBN 9789811285578 (ebook for individuals)

Subjects: LCSH: Plate tectonics. | Geology, Structural.

Classification: LCC QE511.4 .I89 2024 | DDC 551.8--dc23/eng/20231117

LC record available at https://lccn.loc.gov/2023047500

British Library Cataloguing-in-Publication Data

A catalogue record for this book is available from the British Library.

For any available supplementary material, please visit
https://www.worldscientific.com/worldscibooks/10.1142/13657#t=suppl

Desk Editors: Logeshwaran Arumugam/Julio Hong

Typeset by Stallion Press
Email: enquiries@stallionpress.com

Editorial Committee

Editor-in-Chief

Yuzhu Kang
Academician
Sinopec Petroleum Exploration and Production Research Institute
China Petrochemical Corporation, Beijing, China

Involved editors

Shuwen Xing
Institute of Geomechanics
Chinese Academy of Geological Sciences, Beijing, China

Yinsheng Ma
Institute of Geomechanics
Chinese Academy of Geological Sciences, Beijing, China

Yue Zhao
Institute of Geomechanics
Chinese Academy of Geological Sciences, Beijing, China

Dewu Qiao
Institute of Geomechanics
Chinese Academy of Geological Sciences, Beijing, China

Foreword

Since 1970, Yuzhu Kang, a petroleum geologist and geomechanicist, and his team have been studying the "geomechanical theory" originally created by Li Siguang, a famous geologist, and applying the theory to research the effect of controlling oil and gas in the Tarim Basin, Xinjiang Autonomous Region and throughout China. Under Yuzhu Kang's leadership, the team achieved a major breakthrough in the exploration and development of Paleozoic marine oil and gas fields in China in 1984. They discovered the Tahe oil field, an extra-large oil field, in the Tarim Basin in 1990 and subsequently another 20 oil and gas fields in Xinjiang.

Through a repeated process that involved applying insights from theory to practice and then revising theory, the team published more than 10 scientific monographs, including *Oil Control Effects of the Tarim Basin Tectonic System*, *Main Tectonic Systems and Oil and Gas Distribution in China*, *Research on Oil Control Effects in the Junggar Basin, Qaidam Basin, Ordos Basin, Sichuan Basin, Songliao Basin, North China Basin and Other Tectonic Systems* and *Oil and Gas Geomechanics*, and more than 100 papers. Since 2006, they have introduced the theory of geomechanics to the world. After more than 10 years of continuous research and writing, Yuzhu Kang and his team have officially published *Introduction to Global Tectonic Systems*. In this book, a large amount of worldwide geological data is analyzed. The Earth is divided into eight major tectonic systems for the first time, the evolutionary characteristics of all tectonic systems are proposed, the main controlling factors of the formation of tectonic systems are discussed, and innovative views, such as the proposition that no regional metamorphism occurred in the Paleozoic and in some regions in the Meso- and Neoproterozoic eras, are advanced. These innovative theoretical perspectives not only fill the gaps in the understanding of global tectonic systems but also enrich and develop theories of global geoscience. This book is a high-level work with rich content and important theoretical and practical significance.

The efforts of Yuzhu Kang and his team in the further development and innovative advancement of Li Siguang's geomechanics theory are worthy of praise and reading.

Tingdong Li

Preface

Headed by Yuzhu Kang, an academician, the team has carried out geological surveys in many foreign countries since 1985. Especially in the past decade, they participated in the evaluation of comprehensive geological research on global oil and gas resources organized by relevant domestic departments. This book was prepared through systematic study and writing and is based on geomechanical theory. The process involved consulting a large number of monographs, reports and papers published by domestic and foreign experts. The main contents of this book include the following: (1) the origin of Earth's motion is discussed; (2) the main global tectonic systems (E-W-trending tectonic system, N-S-trending tectonic system, N-E-trending tectonic system, N-N-E-trending tectonic system, N-W-trending tectonic system, epsilon-shaped tectonic system, S-shaped or reverse S-shaped tectonic system and rotation and torsion tectonic system) are classified; and (3) the evolutionary features of the tectonic systems and their correlations are specified.

The book was prepared under the leadership of Yuzhu Kang. The main coauthors for the chapters are as follows: the Preface and Chapter 1 were prepared by Yuzhu Kang; Chapter 2 was prepared by Yuzhu Kang, Shuwen Xing, Yue Zhao, Yinsheng Ma and Zongxiu Wang; Chapter 3 was prepared by Yuzhu Kang, Shuwen Xing, Yue Zhao, Yinsheng Ma, Dewu Qiao, Zongxiu Wang, Xingui Zhou, Zhihong Kang and Zhihu Ling; Chapter 4 was prepared by Yuzhu Kang, Zongxiu Wang and Huijun Li; Chapter 5 was prepared by Yuzhu Kang, Shuwen Xing, Zhihong Kang, Yue Zhao, Zhihu Ling, Zhijiang Kang and Huijun Li; Chapter 6 was prepared by Yuzhu Kang, Shuwen Xing, Zongxiu Wang, Zhijiang Kang and

Huijun Li; Chapter 7 was prepared by Yuzhu Kang, Shuwen Xing, Zhihong Kang, Yinsheng Ma, Dewu Qiao, Zongxiu Wang and Zhihu Ling; Chapters 8–10 were prepared by Yuzhu Kang, Shuwen Xing, Zongxiu Wang, Zhihong Kang, Yinsheng Ma, Zhijiang Kang and Zhihu Ling; and Epilogue was prepared by Yuzhu Kang and Shuwen Xing. The whole book was edited and finalized by Yuzhu Kang.

This book was completed under the strong support of many officials from the Ministry of Land and Resources, China Geological Survey Bureau, Southern Marine Science and Engineering Guangdong Laboratory (Guangzhou) and Petroleum Exploration and the Development Research Institute of CPCC, especially with the guidance of Shuwen Xing and Long Changxing, Directors of the Institute of Geomechanics, Chinese Academy of Geological Sciences. Many works published by geologists, oil and gas geologists and other scientific researchers are referenced in this book. I would like to express my heartfelt thanks to these individuals. In the process of writing this book, we have also received the financial support of Key Special Project for Introduced Talents Team of Southern Marine Science and Engineering Guangdong Laboratory (Guangzhou) (GML2019ZD0102), and hereby express our gratitude.

Yuzhu Kang
Beijing
January 2018

Contents

Chapter 1

Introduction

Yuzhu Kang

Abstract

In this chapter, the origin of the Earth's motion, the characteristics of global sea–land changes, the lack of Paleozoic regional metamorphism in the Earth and the major global tectonic movements are introduced. The important driving forces of the formation of tectonic systems include changes in the Earth's rotation speed, the effects of celestial bodies on the Earth, radioactive elements in the crust, the heterogeneity of the crust thickness, differences in the crust density, etc. Under the combined effect of these factors, multidirectional ground stresses were generated in different regions and different eras, resulting in the formation of different tectonic systems. Eight major tectonic system types on Earth are systematically proposed for the first time.

Keywords: Earth; Motion; Tectonic System; Driving Force

The tectonic system is the theoretical core of geomechanics. From the point of view of a global unified movement, this concept regards the deformation of all rocks (including submarine structures) generated by a unified tectonic stress field on the Earth's crust as a whole. In other words, although the various tectonic traces in the Earth's crust differ in shape, size, nature and direction, they are not isolated. Each kind of tectonic trace is bound to be accompanied by some other trace, which together constitute a unified deformation image, i.e., a tectonic system.

The notion of tectonic system is rigorously defined by Mr. Li Siguang as follows: "A tectonic system is a whole system of tectonic zones composed of many tectonic elements with different forms, different properties, different levels, different orders but with genetic relations, in combination with rocks or blocks among them".

The basic premise of attributing all kinds of tectonic traces to a tectonic system is to ensure that there is a genetic relationship among them, i.e., the mechanical connections among these tectonic traces formed in the unified tectonic stress field in the process of formation and development. For example, two sets of shear joints with conjugate relationships have a genetic relation. If they are accompanied by transverse tension fractures and the associated flaser structures or stylolite seams by pressure solution that directly intersected with them, the unity of the stress state reflected by them can be further explained. In geomechanics, this relationship among a set of tectonic traces unified in mechanics is called the generative relation. The unity of the tectonic stress field applies to all components in a tectonic system. For example, in an epsilon-type tectonic system, although the stress state or local stress field is different in the spine, frontal arc and reflection arc, they are mutually restricted and interdependent in the overall stress field. Therefore, there is a generative relationship among them.[1] In a large-scale tectonic system, traces with a generative relation can differ in shape, property, level and order, and they can appear in rocks, blocks and formations of different ages and with different mechanical properties. The local stress state and local stress field reflected in different places can also be different. How, then, can we determine whether there is a generative relation among certain tectonic traces? The solution is based mainly on the following two points. First, although the local stress states or local stress fields reflected by them are different, the overall stress field reflected by them is unified. Second, due to the complexity of geological conditions and the different orders, tectonic traces may appear earlier or later, and the tectonic movement period of their formations must be approximately the same.

Since rock deformation is heterogeneous in a unified tectonic stress field, deformation is strong in some places and weak in other places, and the tectonic system is divided into tectonic zones and blocks based on the relative strength of the tectonic deformation. As these areas are interdependent and inextricably linked, the tectonic system can be described as comprising tectonic zones and blocks.

The main reasons for the formation of the tectonic system include the self-rotation force of the Earth's crust, the effects of celestial bodies on the Earth, the radioactive materials inside the Earth, different thicknesses and densities of various parts in the Earth's crust, etc.

The main types of tectonic systems discussed in this chapter include E-W-extending tectonic systems, N-S-extending tectonic systems, N-E-extending tectonic systems, N-N-E-extending tectonic systems, N-W-extending tectonic systems, epsilon-type tectonic systems, S-shaped or reversal S-shaped tectonic systems and rotational shear tectonic systems.

1.1 The Origin of the Earth's Motion

Mr. Li Siguang, a famous geologist, long ago pointed out that the change in the self-rotation speed of the Earth is an important driving force of its movement. The angular momentum of a rotating object is conserved and generally expressed by the following formula:

$$wI = C$$

where
w: Angular velocity of a rotating object;
I: Moment of inertia of a rotating object around its axis;
C: Constant.

When I changes, w will change in inverse proportion, that is, when I decreases, w increases. If the mass of the Earth moves toward its center, I will decrease. This change may result from several different effects: (1) the whole Earth shrinks (shrinkage theory); (2) large-scale subsidence occurs in the Earth's crust (vertical motion theory); and (3) the gravity differentiation movement and the convection of lava with different densities may take place inside the Earth. Regardless of which of the above hypotheses is closest to reality, as long as any of the effects, including at a certain stage in their development, makes the mass of the earth converge toward its center to a certain extent, the Earth's angular velocity will increase, resulting in an inevitable change in the overall shape of the Earth. On the surface of the Earth or the upper layer of the crust, when the intensity of the resistance against this change is less than that inside the Earth, especially when the isogeothermal surface rises, a horizontal force with a certain intensity is likely to exert a pushing effect on the upper layer of the crust to meet the requirement of the Earth's new shape.

The force generated to create this effect is clearly the horizontal component of the combined action of the increased centrifugal force and gravity by the acceleration in the Earth's angular velocity.[2] This horizontal component can adapt to the requirements for horizontal movement of certain parts of the Earth's crust, especially to form an epsilon-type tectonic system.

Meanwhile, the crust, or the upper layer, is not necessarily uniformly fixed to its base. If the two adjoining parts of the Earth's crust do not accelerate at the same pace as the rotation speed of the Earth, there will be compression and tension cracks in E-W extension. If the part on the east side does not accelerate with the rotation of the Earth as the part on the west side does, compression and torsion cracking will result on the horizontal plane along the N-S extension between the two parts. If the part on the west side does not accelerate with the Earth's rotation as the part on the east side does, tension cracking and torsion cracking will occur on the horizontal plane along the N-S extension. In this case, due to the change in the Earth's angular velocity, the two fractures that strike N-E and N-W can produce not only an E-W extending tectonic system and an epsilon-type tectonic system but also a N-S extending tectonic system.

Based on the principle of the conservation of angular momentum, when the Earth's angular velocity decreases, the moment of inertia around its axis of rotation should increase, that is, its mass distribution should spread outward, and its volume should increase or less dense materials should move upward to the Earth's surface on a large scale. The view that the change in the Earth's moment of inertia causes the change in angular velocity was proposed in China and Hungary (Schmidt) at the same time more than 30 years ago, which cannot be regarded as accidental. A large number of basaltic flows developed during the Permian in southwestern China and other regions of the world. The Deccan dark rocks with an area of approximately $100 \times 104 \text{ km}^2$ exposed in the Indian Peninsula in the early Paleogene, the basic rock flows distributed widely in many regions of the western Indian Ocean, the northern Atlantic and the Pacific Ocean, and various dense igneous rock batholiths and rock masses intruding into the upper part of the crust during various large-scale orogeny eras are all historical traces of the large-scale rise of dense materials below the crust or in the lower part of the crust.

When the Earth's mass distribution changes and its angular velocity decreases due to the continuous influence of the tides, the flatness of the Earth will become too extensive to meet the requirements of the Earth's rotation speed. In this case, some faults and folds with E-W and N-S strikes may occur.

Has, then, the angular velocity of the Earth changed? Affirmative answers can be obtained from the records of ancient solar eclipses and the observations of some astronomers in modern times. Most of these astronomers believed that the angular velocity of the Earth tended to slow down, but others (such as Yurie) held the opposite view. In fact, historical records indicated that the Earth's rotation speed was sometimes slow and sometimes fast. These changes were irregular. Historically, although such irregular changes were not large, they were not used to measure the possible historical changes in geological ages. That is, we have no reason to rule out the possibility that the accumulated effects of the changes in the Earth's rotation speed might sometimes exceed the critical value for the Earth to maintain its surface shape in geological ages.[3] Therefore, we have no reason to conclude that the concentration of the Earth's mass stopped before the movements of the crustal surface of the Earth reached the critical state.

1.1.1 *Effects of Celestial Bodies on Earth*

The reasons for the change in the Earth's angular velocity and the conditions of the gradual directional movement in and under the Earth's crust may relate to the celestial bodies closely related to the Earth, especially the Moon and the Sun, or from the inside of the Earth itself. Let us discuss the possibility of the influence of celestial bodies first. From the perspective of the requirements for the directional movement of the crust, some astrophysicists (such as Taylor, Jolie, Lichkov, etc.) thought that the tidal effects of the moon on the Earth are the general reason for tectonic movement in the Earth's crust. Several astronomers and astronomical geologists from the former Soviet Union argued that the Sun's activities might affect movement in the crust. Among them, some, such as Egenson, believed that the Sun's activities might affect the angular velocity of the Earth, while others, such as Snarsky, believed that the Sun's activities are related to changes in the intensity of the Earth's magnetic field. Based on the hypothesis that when the magnetic field weakens, substances freed from the magnetic pull generate heat and the fact that the 11-year changing cycle of the Earth's magnetic field corresponds to the 11-year cycle of solar activities, Snarsky proposed that the rise and fall of the isogeothermal surface of the Earth's crust was not the cause but the consequence of the change in the magnetic field intensity. Snarsky's novel hypothesis proposed how changes in the intensity of Earth's magnetic field created the conditions for making tectonic movement possible in the crust.

1.1.2 Reasons from the Inside of the Earth Itself

Many radioactive substances are widely distributed in the Earth's crust. These substances constantly generate heat. The temperature at the surface of the Earth remains approximately unchanged. However, the temperature below the crust gradually increases under conditions of the heat transfer rate of rocks and the fixed geothermal gradient. Based on this fact, Jolie concluded that the rocks below the Earth's crust would melt approximately once every 30 million years. Schmidt's "On the Origin of the Earth's Crust" further developed the importance of radioactive substances to the thermal history of the Earth. Many other geologists contributed many works to this field, including Holmes, who used the radioactivity of minerals to identify the Earth's age in the early stage. The existence of radioactive elements in the Earth's crust seems undoubtedly to be closely related with the Earth's thermal state. However, how radioactive elements are distributed in various parts of the Earth's crust and below the Earth's crust remains an open question. It is not reliable to determine the distribution law of the radioactive substances in and below the Earth's crust based on only the radioactivity of some types of rock specimens. As pointed out by Kraskovsky, in the past, much of the data gathered on the thermal state of the Earth in various places were unreliable.[4] In this case, the degree of change in the isogeothermal surface proposed by Jolie and other geologists based on the assumption that the radioactive elements are distributed in a certain regularity in the crust needs to be closely evaluated and investigated. This effect creates conditions for the directional movement of the Earth's crust. Therefore, the temperature difference inside the Earth may cause local movement in the Earth's crust.

1.1.3 Ground Stress Caused by Differences in Crustal Thickness and Density

During the rotation of the Earth, compressive and tensile stresses form due to the different thicknesses and densities of the Earth's crust, which may cause various tectonic deformations. Additionally, regions with greater crustal thickness and higher rock density produce compressive stress in regions with thinner crustal thickness and lower rock density, resulting in corresponding deformation.

1.1.4 *The Continents Could Not Drift in the History of Geological Evolution*

The continental drift hypothesis put forward after being repackaged by Wegener in the early 1920s was dominant in the past and is still believed by some individuals.

Based on the generation of stress from the Earth's movement, the characteristics of crustal movement, and the tectonic movements, major earthquakes, seawater advances and retreats and volcanic eruptions that have occurred on the Earth's crust in the past million years, it is believed that there is no large-scale drift but only ascent and subsidence, strike-slip and subduction effects caused by deep and large faults under horizontal compressive forces in all land blocks on the Earth. Especially when tensional deep and large faults are cracked, magma in the mantle will invade or erupt to seal faults such as gels. The continental blocks changed numerous times in their geological periods. These changes are not the large-distance displacements of continental blocks or continental drifts but changes in the ocean. They are changes in the extents of submergence and exposure of the continents caused by changes in the advance and retreat of seawater. In other words, changes in the ocean bring about changes in land blocks. For example, much land was submerged during the large-scale tsunami generated by the Fukushima earthquake in 2011, resulting in reduced land area on the Japanese island, which was not the drift of the island. Therefore, it can be inferred that changes in and displacements of the continents in geological history resulted mainly from changes in sea and land.[5]

The deep and large faults caused by ground stress have the following characteristics: (1) squeezing, i.e., one block can subduct under another; (2) creation of mountains under the compression effect (uplift); and (3) strike-slip faults, i.e., two fractured blocks may slide away from each other for several kilometers. However, these relative movements are just local displacements on the Earth. In a certain historical period, changes in the ocean were the greatest driver of changes in land blocks, such as changes in various continents in different periods. Due to the above reasons, the various types of tectonic systems formed in the crust have unique characteristics.

1.2 Characteristics of Global Land–Sea Changes

Based on analyses of the global deposition rock distribution, the state of tectonic movement and denudation at different times, it is believed that the land area was relatively large in the Sinian period in the Early Paleozoic Cambrian, and the land area was smallest while the ocean area was largest. The land area in the Ordovician was larger than that in the Cambrian. The land area in the Silurian was the largest in the Paleozoic epoch. The land area in the Devonian in the Late Paleozoic was relatively small, while it was largest in the Permian. Since the Mesozoic, the land area has increased, and the seawater area has decreased continuously. In the Paleogene, the prototypes of the current continents were preliminarily formed.

The land–sea changes are definitely not the result of continental drift but the result of the effects of the following factors:

(1) Under the compression effect of crustal stress, some parts of the crust rose, while some other parts subsided. The seawater flowed into the subsidence zone, and the uplifted parts became land. Crustal uplift and subsidence were constantly and unevenly occurring, leading to the corresponding advance and retreat of seawater.

(2) Factors affecting the advance and retreat of global seawater also include the reduction in seawater area and the increase in land area during the global ice age, such as the Late Sinian, Late Ordovician–Early Silurian, Late Carboniferous–Early Permian and Early Quaternary.

(3) Under the action of ground stress, many multidirectional large-scale faults were produced in the Earth's crust, which promoted the relative movement of the Earth's crust, such as relative uplift or subsidence, or relative translation, and relative uplift of the land blocks.

1.3 No Regional Metamorphism in the Paleozoic

In recent years, the geological investigation of major orogenic zones and large basins to explore oil and gas resources in China and other parts of the world has uncovered that there was no regional metamorphism in the Paleozoic, in contrast to what was originally believed by some experts and scholars. Some examples of orogenic zones investigated in China include those of the Daxinganling Mountains and Changbai Mountains of Northeast China and those of Altay in Xinjiang, Tianshan, the Kunlun Mountains, the Altun Mountains, and the Qilian Mountains in Qinling. The Paleozoic depositional rocks experienced local dynamics or thermal metamorphism only in the vicinity of large fault zones and magmatic rock contact zones. These dynamic and thermal metamorphism zones have a general width of 5–550 m, with the largest width spanning thousands of meters. However, no metamorphism has been found away from fault zones and contact zones. The lithologies with dynamic metamorphism are different, and they are mostly phyllite and slate, as well as schist.[6]

There is no metamorphism in the Paleozoic in the large deposition basins in China (Songliao Basin, Bohai Bay Basin, South Huabei Basin, Ordos Basin, Sichuan Basin, Junggar Basin, Tarim Basin, Qaidam Basin, Corridor Basin, Jianghan Basin, etc.).

There are oceanic facies and oceanic-continental facies carbonate and clastic rock sediments in the Paleozoic in the Eastern and Western Siberian Basins of Russia, equivalent to Lifel in the Sinian, and no metamorphism was found.

There is also no metamorphism in the Paleozoic in the major large basins in the United States, Canada and Brazil on the North American continent.

1.4 Major Tectonic Movements on Earth

The tectonic evolution of the continental lithosphere is caused mainly by tectonic movements and tectonic systems. Their formation, evolution and distribution are restricted by crustal movements, tectonic systems and trends. Therefore, the natures, periods and change laws of the main crustal movements can be discussed in terms of the formation and evolution of the continental crust structure.

Crustal movements can be orogenic, oscillatory, etc., with many types and scales. A tectonic movement in the same period can show different natures under different geological backgrounds. For example, a movement might create mountains in some regions due to compression or strike-slip uplifting and continuous subsidence due to depression or tensioning in other regions, making the later sedimentation behave as unconformity contacts in the former and continuous sedimentation without obvious discontinuity in the latter because compression and tensioning as well as fold uplift and depression (or fault depression) in the crustal movement have always been complementary or accompanied each other. However, because many different products were generated by the tectonic deformation of compression and tensioning caused by tectonic movements, the compression or compression shear deformation zones were often accompanied by deformation events and metamorphism and magma intrusion activities; extensional rift zones were often accompanied by volcanism, such as magma eruption, forming active ocean troughs and volcanism, i.e., clastic rock formations; and depression or oscillation zones were often accompanied by relatively stable crushed rocks, i.e., carbonate rock formations, playing an important role in controlling deposition minerals or volcanic deposition minerals. The original structure patterns could be changed by later tectonic activities, showing obvious novelty, or could be inherited partially, showing obvious inheritance, which was related to the boundary conditions and the distribution and variation in the tectonic stress field of the affected parts during the tectonic movement period (as shown in Table 1.1).

Tectonic evolution histories experienced by the major blocks in China are not identical. The manifestations and strengths of tectonic movements are quite different in different periods and blocks. The basic situation is as follows: the North China block was consolidated in two tectonic movements at the ends of the Neoarchean and Paleoproterozoic, forming a crystalline basement, which was in a stable caprock development stage in the Mesoproterozoic and showed deformation and metamorphism with greater activity intensity in the Indosinian movement in the Late Triassic. The Yangtze–Talimu block was consolidated to form a crystalline basement at the end of the Paleoproterozoic, which was in an active and subactive deposition environment in the Meso-New Proterozoic, to form the fold basement in the Jinning–Chengjiang Movement (called the Altyn–Talimu Movement in Xinjiang). The block began to enter the stable caprock development stage in the Danian. The development process of this tectonic block has been basically consistent since the Phanerozoic, except that the intensity of activities was sometimes greater in the west. However, since the Indosinian movement, there has been a clear difference between the east and the west. The ancient Cathaysia block was consolidated to form a crystalline basement at the end of the

Table 1.1.   Comparison of geological ages and geological motions.

Geological Age		Isotope Age Value/Ma	Tectonic stage and crustal movement			
			Major Geological Event	Europe and America	China	Africa
Quaternary	Holocene	现在				
	Pleistocene	0.01			Himalayan Movement (Late) — Himalayan stage	Late Alps Movement — Alps stage
Neogene	Pliocene	2	United paleoland disintegration stage			
	Miocene	5				
Paleogene	Oligocene	22.5		Saaf Movement — New Alps stage	Himalayan Movement (Early)	
	Eocene	37.5		Pyrenees Movement		
	Paleocene	50				Early Alps Movement
Cretaceous		65		Laramie Movement — Old Alps stage	Yanshan Movement (Late) — Yanshan stage	
		137		New Westville Movement	Yanshan Movement (Mid)	
Jurassic		185		Old Westville Movement	Yanshan Movement (Early) / Indochina Movement (Late)	Third Act of Hercynian Movement — Hercynian stage
Triassic		230		Appalachian Movement — Hercynian stage	Indochina Movement (Early) — Indochina stage	Second Act of Hercynian Movement
Permian		280			Yiming Movement — Hercynian stage	
Carboniferous		350	United paleoland formation stage	Breton Movement	Tianshan Movement	First Act of Hercynian Movement
Devonian		400				
Silurian		440		Iri Movement	Qilian (Guangxi) Movement — Caledonian stage	
Ordovician		500		Taikang Movement — Caledonian stage	Gulang Movement	(Katanga) Late Pan-African Movement — Pan-African stage
Cambrian		610			Xingkai Movement	
Sinian		850		Anetti Movement	Jinning Movement (Late) — LvLiang/Jinning stage	
New		1055	Land platform formation stage	Goethe-Greenwell Movement	Jinning Movement (Early)	(Katanga) Early Pan-African Movement
Mid		1600–1700		Carrie-Hudson Movement	LvLiang (zhongtiao) Movement — Fuping/LvLiang stage	
New		2500–2600			Wutai Movement	
		3000–3900	Land nucleus formation stage	Sam-Kennoll Movement	Fuping Movement	
		3800				
Hades		4600	Astronomical stage			

Paleoproterozoic, which was in an active depositional environment from the Mesoproterozoic to the Early Paleozoic, i.e., an active deposit before and after the Jinning Movement. The fold basement was formed in the Paleozoic and then transformed into a stable caprock deposit. It has shown strong activity since the Indosinian period. The southern Tibet–western Yunnan region belongs to the northern margin of the Gondwana continent and is similar to the Cathaysia block. Its fold basement was formed in the early period of the Early Paleozoic and evolved into the main component of the East Tethys tectonic domain with strong activity in the Mesozoic.[7] The northernmost region of China belongs to the southern margin of the Mongolian block, in which the basement rock is rarely exposed. Existing data indicate that there was a vast ocean in this region during the Paleozoic epoch. At the end of the Paleozoic, the folds returned and entered a stable development stage. It is possible that there was a Proterozoic crystalline basement. However, in the Jiamusi-Laoyeling, Eastern Jilin and Eastern Liaoning regions, the blocks uplifted for a long period of time in the Mesoproterozoic and strongly subsided in the Neoproterozoic, which is consistent with the conditions in the Dabie–Jiaodong region and similar to the Yangtze–Cathaysia block since the Neoproterozoic but noticeably different from the North China block.

The following sections briefly introduce the main tectonic movements in the Paleozoic are briefly introduced in order from old to new.

1.4.1 *Tectonic Movements Between the Mesoproterozoic and the Cenozoic*

The Jinning Movement originally referred to an important tectonic movement with deformation, metamorphism and fold basement formation in the Kunyang Group and the Huili Group in Southwest China. Later research discovered that this movement was widespread in the Yangtze block and its periphery and created an important deformation and metamorphism to form the basement in this region. In the later period of this block, there was extensive magmatic activity and transitional active formations, so the subsequent Chengjiang Movement, the tail act of the Jinning Movement, or tectonic changes between the Sinian glacial rock series and the underlying rock series should be included with the time limit of 800–1000 Ma. This tectonic movement manifested strongly in the Tarim–Qaidam block, Songpan block in northwest Sichuan, Qiangtang block in northern Tibet and the uplift zone from the south of the Alxa block to the south of the North China block and the Funiu mountainous region. It was called the Altyn Movement and Tarim Movement or Quanji Movement in Northwest China. There were two interfaces in many regions, with the main interface at approximately 1000 Ma and a later interface at approximately 800 Ma, which were approximately equivalent to the bottom and top interfaces, respectively, in the Qingbaikou period. The deposition formation, tectonic deformation and metamorphism and the magmatic activities of the fold basement in the Yangtze–Talimu block were basically comparable. In addition, the Qingbaikou volcanic-clastic formation and flysch-like formation in these regions were basically formed in the same tectonic environment, and the overlying Sinian glacial rock and carbonate rock formation and the late phosphorus (vanadium, uranium) formation were more consistent. In general, in the Jinning–Talimu movement, the basement folds of the Yangtze–Talimu block were consolidated and transformed into the stable caprock development stage. This movement also had an important impact on the North China block, which manifested mainly in the uplift in the Sinian and unaccepted sedimentation during 600–800 Ma. The Yangtze–Talimu block and the North China block formed a stable unified continental crust in early China. The deformation domain of the Jinning–Talimu movement was involved in the southwestern North China block. Along the Beishan–South Alashan–Funiushan line, the Meso-Neoproterozoic Erathem in the Pre-Sinian was significantly deformed and metamorphosed and became the northern edge of the Yangtze–Talimu block, followed by the Sinian deposition area. This zone might extend northward to the eastern margin of the Junggar Basin.

1.4.2 *Tectonic Movement Between the Silurian and Devonian (The First Act of the Hercynian Movement)*

This tectonic movement was an important tectonic movement at the end of the Early Paleozoic and was widespread on all continental crusts. However, intensities, manifestations and time limits of occurrence and ending differed among land blocks or even in different regions of the same block. Therefore, the names of different land blocks and different regions are not consistent. The most representative events are the Qilian Movement in Northwest China and the Guangxi Movement in South China, which formed from fold deformation in the late Early Paleozoic.

The North China block was created by the overall uplift and denudation from the Late Ordovician to the Early Carboniferous, without the upper Ordovician–lower Carboniferous. There was a regional parallel unconformity contact between the middle Carboniferous sediments and the underlying middle Ordovician.

In some active zones along the periphery and inside of the Cathaysia block and the Yangtze block, fold movement occurred at the end of the Silurian, resulting in strong deformation of Lower Paleozoic rocks, accompanied by varying degrees of dynamic metamorphism, medium-acidity magma intrusion and significant fault activity and fault block uplift. For example, tight linear folds and large fault zones formed in the North Qinling Zone, the Longmen Mountain–Yulong Snow Mountain Zone, the eastern margin of the lower Yangtze and the Cathaysia Block. Significant unconformity contacts are evident between the Devonian and Lower Paleozoic in the Ailao Mountain Zone on the western and southwestern margins of the Yangtze Block and Ziyun-Luodian fault fold zone. The Sichuan Basin was uplifted, and the adjacent Daba Mountain and Dalou Mountain were also generally uplifted but without significant deformation. The middle and lower Devonian and Silurian were mostly eroded unconformities. The lower Yangtze block was in a similar condition.

1.4.3 *Tianshan Movement (The Second Act of the Hercynian Movement)*

The Tianshan movement was an important tectonic movement in the middle and late periods of the Late Paleozoic. Due to the strong reformation of the Chinese continental crust by the Caledonian movement, the Cathaysia block and the Yangtze block became integrated, resulting in a change in the boundary conditions between eastern and southeastern China. In addition, since the Indian block was close to the Tarim–Yangtze block and the Caledonian arc tectonic zone in Mongolia formed and developed, the pattern of the continental crustal structure in China has gradually changed since the Late Paleozoic. In the early period of the Late Paleozoic, the overall continental crust in China changed from uplift to stable deposition. There were inherited activities in some Caledonian active zones, with frequent ups and downs. The Devonian system was composed of mainly piedmont-intermountain continental or continental-littoral shelf facies clastic rocks, basically without normal marine carbonate formation and even missing in many areas.[8] In the late Hercynian movement, these rocks formed mainly between the Early and Late Permian. The former manifested by fold deformation and magmatic activity with significant unconformities, while the latter manifested mainly by uplift discontinuities with local unconformities. On the North China block, they only manifested by weak discontinuities. Although the manifestations varied from place to place, they generally reflected the extensive influence of the Tianshan movement on the continental crust in China. This movement showed that two zones formed in the middle and late periods of the Late Paleozoic. After many activities, they gradually formed in W-E extension at the beginning of the Late Permian, and a fold zone was formed and accompanied by strong magma intrusion after deformation and then was unconformably covered by the stable clastic-carbonate rock formation in the Late Permian. During this period, there were also several near-NS-trending plateau basalt zones, especially the formation of the Emeishan basalt zone along the north–south zone of Sichuan–Yunnan, the basic intrusive rock zone containing vanadium–titanium magnetite in the Panzhihua–Xichang region, and the tectonic magma dynamic metamorphic zone in the Jinsha River–Lincang region. Notably, the tectonic movement of the continental crust in China in the Late Paleozoic did not end between the Late Permian and Triassic in many areas but between the Early and Late Permian, while there was continuous deposition without unconformity in the Late Permian–Early Triassic.

1.4.4 *Tectonic Movement in the Middle and Late Triassic*

Strong tectonic movement occurred in the Triassic. Its main act occurred in the Middle and Late Triassic, or the middle and late period of the Late Triassic. This movement brought the evolution of the tectonic system between the Asian continent and the Pacific block into a new stage. There were strong and obvious traces in the continental crust in China, the East Asian Pacific region, and the Indosinian region, especially the development and formation of the two E-W-trending tectonic systems throughout central China, the rise of the Tethyan tectonic zone in southwestern China, the formation of the NE-trending structure in eastern China and the formation of the Indosinian active tectonic zone on the southern margin of the continent, which caused Late Triassic Retician-Early Jurassic Riasian sediments to be widely and unconformably distributed on the metamorphic rocks strongly deformed in the Middle and Late Triassic. The western Tethys region evolved into an active marine deposition environment, and the marine invasion ended and turned into postoro-genic continental coal deposits in the remainder of China. After the redeformation in the Indosinian tail at the end of the Late Triassic, seawater in the southwest continued retreating westward to a corner of southern Tibet, accompanied by medium acidity magma intrusions and tectonic dynamic metamorphism, which developed extensively in the Indosinian movement.

After experiencing the strong and extensive Indosinian movement, the present unified Chinese continental crust was basically formed, and the boundary conditions among the three major tectonic domains were set. The subsequent Yanshan Movement and Himalayan Movement were mainly reformed and developed on the basis of the Indosinian framework, with only some local or regional changes. Therefore, the Indosinian movement represented another major change in the evolution of the continental crust in China is comparable to the Luliang Movement and the Jinning Movement.

1.4.5 *Tectonic Movement Between the Jurassic and the Cretaceous*

The Yanshan Movement is generally considered an important tectonic movement widely developed throughout China during the Jurassic–Cretaceous period and manifested mainly as fold-fault changes, magma intrusions and eruptions, and some local metamorphisms. It not only was an important tectonic movement in the continental crust in China but also had an important impact on the Peri-Pacific region and the middle Tethys tectonic domain. Due to differences in manifestation and deformation intensities of this movement across various regions, there was a divergence in the division of the tectonic periods. These periods were generally divided into three strong fold-fault deformation periods and two weak deformation periods (five acts in total: Middle–Late Jurassic, Late Jurassic–Early Cretaceous, Cretaceous, Late Cretaceous-Palaeogene). The magmatic activity and tectonic deformation in the Late Jurassic–Early Cretaceous were the most obvious. However, there was still a large divergence in the movement characteristics and the methods of manifestation between the Late Cretaceous and Paleogene because the depositional environment between the Late Cretaceous and Paleogene was basically the same, and the strata were mostly continuous without obvious discontinuities and regional unconformities.

The Yanshan Movement in China was actually the continuation and development of the Indosinian Movement, which changed from the north–south balanced compression in the Hercynian-Indosinian period to the non-equilibrium compression and torsion environment. In the Yanshan period, it was dominated by non-equilibrium torsion. The deformation characteristics changed from strong plasticity to brittle plastic deformation. The Yanshan Movement not only created new tectonic patterns but also strengthened and inherited some early tectonic types and tectonic patterns, having formed the tectonic features of the present continental crust in China.[9]

1.4.6 *Tectonic Movement Between the Paleogene and the Neogene*

This tectonic movement has occurred in the continental crust in China since the Cenozoic. It turned the Mesozoic Tethys Sea area into a large mountain range and caused the formation and development of the trench arc basin near the Pacific Ocean. The main deformation and metamorphism events occurred from the beginning of the Oligocene to the beginning of the Miocene. From the beginning of the Oligocene, the seawater in Tibet completely retreated, followed by violent deformation and metamorphism, manifested by strong folds and faults, medium-acidity magma intrusion, and dynamic thermal metamorphism. In the later stage, large-scale thrust, nappe and sliding stacking structures were formed, resulting in large-scale uplift and an east–west extension of the Tibetan Plateau crust, which triggered the lateral migration of substances, causing the eastern crust to twist clockwise around the Tibetan block. In this way, the Pamir–Himalayan region became the highest and youngest fold mountain system in the world. The Eastern China region was restricted by the Pacific tectonic domain. In the Neogene and early Quaternary, a nearly north–south trending fold mountain system was formed in Chinese Taiwan and the Philippines. The NE-NNE-trending tectonic zone and the East Asian island arc zone in the continental shelf of southern China were further developed.

The first act of the Himalayan movement occurred between the Neogene and the Paleogene, resulting in strong unconformity contacts between them. The second act occurred in the southeastern waters and the Chinese Taiwan–Philippines region between the Miocene and the Pleistocene, known as the Taiwan Movement. It reached a stage of violent activity before the Pleistocene. The Pleistocene was strongly unconformable over the Pliocene. There were not only fold deformations but also high-pressure dynamic metamorphic rock zones. The third act occurred from the Pleistocene to the present. The western plateau was uplifted sharply under the actions of SN-trending compression and unbalanced torsion stress, and the old faults were active again. Quaternary volcanic eruptions occurred in some areas, and there was more significant strike-slip activity on the eastern margin. Under the action of the Pacific block, the eastern area was subjected to east–west compression, resulting in right-side strike-slip along the early NE-NNE-trending fault zone, forming a series of secondary pull-apart basins and changing the mechanical properties and movement modes of the early faults.[10] Thus far, its activity is still very strong, which is crucial for controlling the seismic activities and other geological disasters in the region.

In this book, 8 types of major tectonic systems in the world have been systematically established for the first time: (1) E-W-trending tectonic systems; (2) N-S-trending tectonic systems; (3) NE-trending tectonic systems; (4) NNE-trending tectonic systems; (5) NW-trending tectonic systems; (6) epsilon-shaped tectonic systems; (7) S-shaped or reverse S-shaped tectonic systems; and (8) rotation and torsion tectonic systems, with E-W-trending and N-S-trending tectonic systems as the main systems.

The strengthened characteristics of all tectonic systems are given below: stages, inheritance, difference, migration and transformation. These characteristics show the complexity of all tectonic systems. The genetic evolution of various large and small blocks is controlled by tectonic systems, while the formation and evolution of tectonic systems are controlled and affected by each block. The interaction between them resulted in the present global tectonic patterns and the changes in the sea and land blocks.

The formation and evolution of tectonic systems controlled the formation of deposits and prototype basins at various ages as well as the formation, reformation and finalization of the energy mineral resources and metallic mineral resources. Moreover, the tectonic systems' control of the distribution of mineral resources is quite regular.

This book is a pioneering and innovative masterpiece and a major contribution to global geology.

References

1. Li Siguang. *Introduction to Geological Mechanics*. Beijing: Sciences Press, 1973.
2. Li Siguang. Vortex Structure and Other Complex Problems Related to Geotectonic System in Northwest China. *Acta Geologica Sinica*, 1954, 34: 339–410.
3. Aadland, R. K., Schamel, S. Mesozoic Evolution of the Northeast African Shelf Margin, Libya and Egypt. *AAPG Bulletin*, 1988, 72(8): 982.
4. Abdel, A., Shallow, J. A., Nada, H. *et al*. Geological Evolution of the Nile Delta, Egypt, Using REGL, Regional Seismic Interpretation. *Proceedings of 13th Petroleum Exploration Conference*. Egypt: Egyptian General Petroleum Corporation, 1996: 242–255.
5. Acharya, S. K. Mobile Belts of the Burma–Malaya and the Himalaya and Their Implications of Gondvana and Chthaysia/Laurasia Continent Configurations. *Proceedings of the Third Regional Conference on Geology and Mineral Resources of Southeast Asia*. Bangkok, 1978: 121–127.
6. Agrawal, B., Fish, S. F. Trace Metals in Crude Oils from Assam Oilfield, India. *Indian Journal of Technology*, 1972, 10(3): 117–119.
7. Alexander, E. M., Pegum, D., Tingate, P. *et al*. Petroleum Potential of the Eringa Trough in South Australia and the Northern Territory. *Journal of the Australian Petroleum Production and Exploration Association*, 1996, 36(1): 322–349.
8. Li Siguang. *Geomechanical Technique*. Beijing: China Sciences Publishing & Media Ltd. (CSPM), 1976.
9. Li Shujing *et al*. *Introduction to Classification and Characteristics of Major Tectonic Systems in China*. Institute of Geomechanics, Ministry of Geology and Mineral Resources. Collection of Studies on Provincial Tectonic Systems in China (Volume 1). Beijing: Geological Publishing House, 1985.
10. Li Siguang. *Abstract of Astronomical, Geological and Paleontological Data*. Beijing: Science Press, 1972.

Chapter 2

E-W-Trending Tectonic Systems

**Yuzhu Kang, Shuwen Xing, Yue Zhao, Yinsheng Ma,
and Zongxiu Wang**

Abstract

In this chapter, E-W-trending tectonic systems are introduced, including those in the Arctic region, China, the Northern Hemisphere and the Southern Hemisphere.

Keywords: Tectonic system; E-W-trending; Type

The E-W-trending tectonic system is a global system that is distributed along certain latitude lines around the Earth. According to the current data, there are strong compressional structures, magmatic activity and metamorphic zones at approximately 8°–10° intervals along latitude lines. They are composed of mainly various E-W trending fold zones, compressional fault zones and magmatic rocks, obliquely intersected with torsional faults and vertically accompanied by tensional faults. In the zones, there are generally some inclusive or merged ancient land blocks or rock blocks and some intermittent E-W-trending troughs or basins. Due to their large scale, they affect the crust at greater depths. There are not only extensively developed medium-acidic magma rocks but also a large amount of mafic and ultramafic magma rock intermittently distributed along the tectonic zones. In addition, due to extensive and strong plastic-brittle deformation, there are often large tectonic dynamic metamorphic rock zones, such as low-temperature, high-pressure metamorphic rock zones, large ductile shear zones, migmatites, remelted granite zones, etc. Each tectonic zone has its own system. They extend along a certain latitude, cross continents and oceans, and are widely distributed on Earth. As a system, all tectonic zones extend into lands and oceans at certain latitudes and are widely distributed on Earth. Due to interference from other tectonics and N-S-trending slip or deflection of blocks, the various zones and sections are currently located differently in latitudes and are not necessarily all E-W trending. The earlier the tectonics formed, the larger the degree of slip or deflection is. Some sections are often inclusive in the newer E-W-trending tectonic zones. Most of these giant tectonic systems formed in the Late Paleozoic–Mesozoic and Cenozoic eras. Some of them have their inheritances, showing a wavy or sometimes sinusoidal shape in trending. Since some sections slip along the E-W-trending fault zones (strike-slip) or the N-S-trending fault zones, several secondary derivative torsional tectonic systems were often generated in the zones. The above-mentioned complex factors made the tectonic plane more complicated, so it is known as the giant E-W-trending complex tectonic zone. The formation of tectonic systems was restricted by the E-W-trending concordant function band produced by the rotation of the Earth, so they have distinct orientation and positioning features. They all have the characteristics of worldwide distribution and are generally distributed at an interval of latitude. Generally, a strong compressional tectonic zone appears at 8°–10° intervals in the middle and high latitudes of the Earth, which is consistent with the critical latitude.[1]

The E-W-trending tectonic systems are summarized in order from north to south in the following sections.

2.1 E-W-Trending Tectonic Systems in the Arctic Region

A series of nearly E-W-trending basins at 70°N around the Arctic developed, such as the West Siberia Basin, East Siberia Basin, Rattev Basin, Kekima Basin, East Siberian Sea Arctic Slope Basin and Victoria Basin (Fig. 2.1). The orogenic zones, fault zones and magmatic rock zones are scattered along these basin margins in an E-W-trending direction. The known tectonic patterns within these basins are also distributed in a nearly E-W-trending direction.

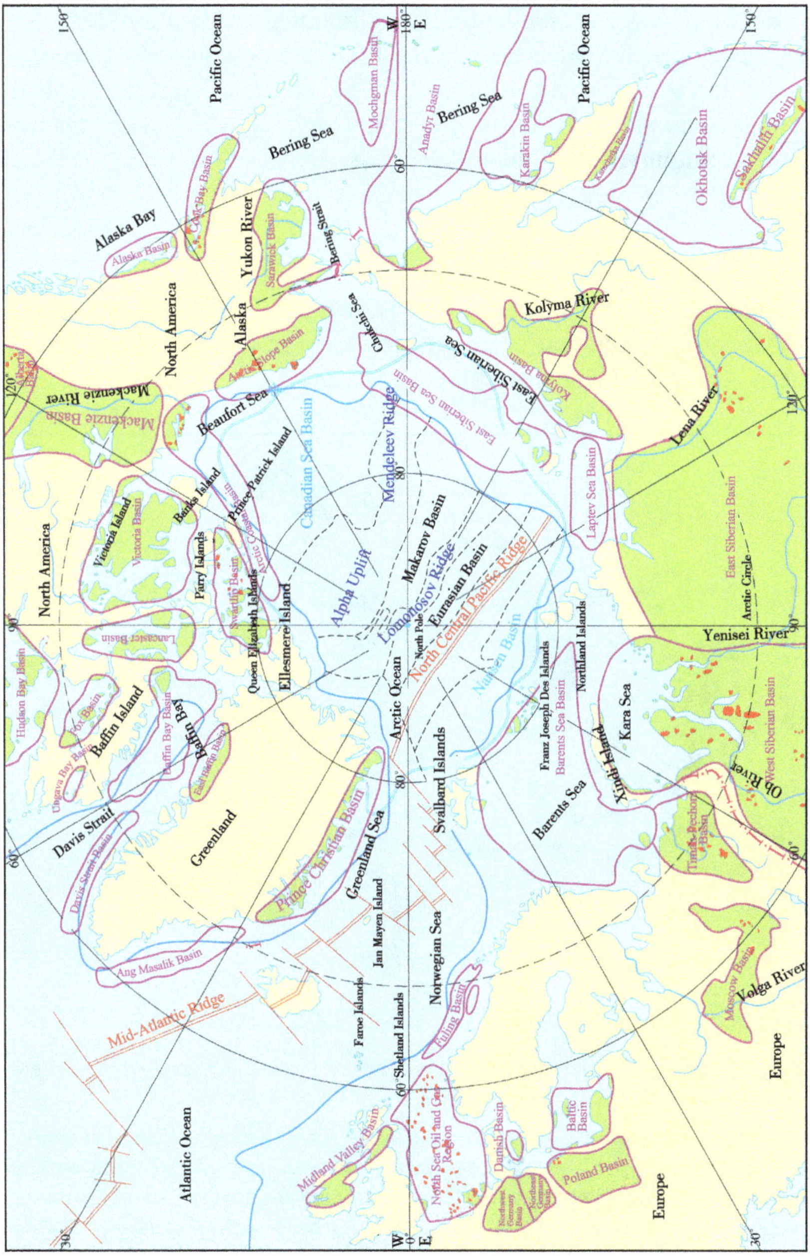

Fig. 2.1. E-W-trending tectonic systems in the Arctic region.

2.2 E-W-Trending Tectonic Systems in China

In order from north to south, the E-W-trending tectonic systems that clearly developed in China include the Ilhuli tectonic system, Yinshan–Tianshan tectonic system, Kunlun–Qinling tectonic system, Nanling tectonic system and Xisha tectonic system (Fig. 2.2). Their main characteristics are summarized in the following sections.

2.2.1 *E-W-Trending Tectonic System in the Ilhuli Region*

This zone is distributed at 50°N–52°N, showing more obvious features in Ilhuli Mountain to the east of the greater Xing'an Mountains and in the northern section of the smaller Xing'an Mountains.

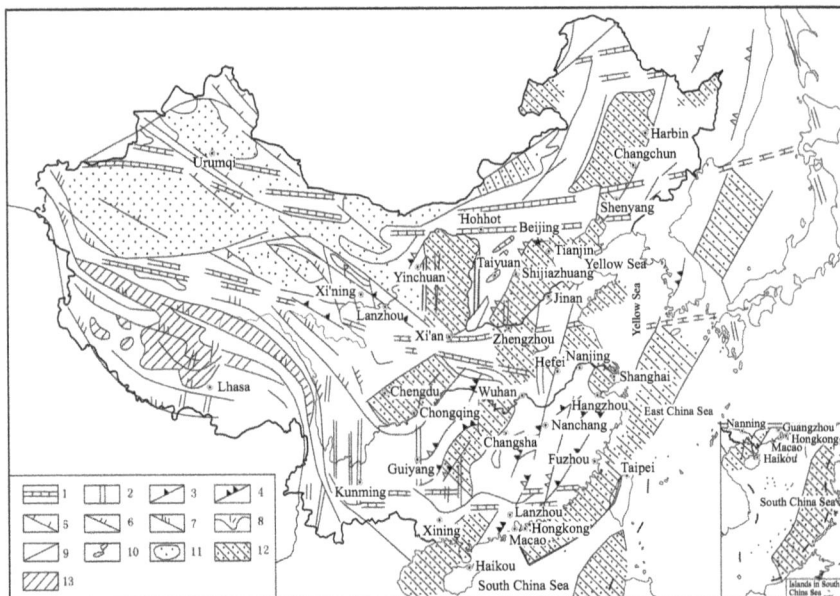

Fig. 2.2. Distribution of the main tectonic systems and the large Paleozoic oil and gas fields in China (from Sun Dianqing, modified).

Notes: (1) E-W-trending tectonic system; (2) N-S-trending tectonic system; (3) Cathaysia system; (4) New Cathaysia system; (5) Xiyu tectonic system; (6) Hexi tectonic system; (7) Reverse S-shaped tectonic system in Qinghai, Tibet, Sichuan and Yunnan; (8) Epsilon-shaped and arc-shaped tectonic system; (9) Other tectonic system; (10) Multi-font-shaped basin; (11) Oil field controlled by the western region tectonic system; (12) Oil field controlled by the new Cathaysia tectonic system; (13) Oil field controlled by the Pamir–Himalayan reverse S-shaped tectonic system.

It is generally a compound uplift zone, including the E-W-trending Proterozoic and Upper Paleozoic compound folds, Jinning and Hercynian granite zones and several associated large faults. The Cretaceous volcanic rock zone is also involved, with uplift geomorphology.[2] This zone might have been generated in the late Paleozoic or earlier. There were some activities in the Yanshanian period, and there is still modern crustal movement. This zone was cut and destroyed by the N-N-E-trending first-order uplift zone and subsidence zone, so it has poor continuity.

In the western margin of the zone, the nearly E-W-trending Tangnu–Kent tectonic zone traces the border between Russia and Mongolia, which is approximately equivalent to the Almoriga fold zone that traverses Belgium. This Paleozoic fold zone might have developed considerably along the Newfern Seawall on the seabed plateau in the North Atlantic Ocean. In Quebec and Ontario of Canada, geophysical investigation has revealed an E-W-trending anomalous gravity zone, which might be a trace of this zone. Therefore, this zone appears to be part of the global E-W-trending tectonic system.[3]

2.2.2 *Yinshan–Tianshan E-W-Trending Tectonic System*

The main body of this tectonic system is located at approximately 40°N–43°N, with wider or narrower spreads in some regions and some fluctuations and deflections in directions. It stretches for approximately 4,000 km in China and is a very significant tectonic zone that traverses northern China. Its geomorphology is quite distinct. It has played an important controlling role in geological history. Its western section includes the entire Tianshan Mountains, the Alay Mountains of Tajikistan and the Kyrgyz Mountains of Kyrgyzstan. Further eastward, it extends to the North Mountain of Gansu, covered by the Badain Jaran Desert. It connects with the middle section of Yabryan Mountain after passing through the north side of the mountain. It spreads approximately along the south side of the border between China and Mongolia. After passing through Langshan, Baiyun Obo, Yinshan, Daqingshan and the Dama Mountains, it connects with the Yanshan Mountains. Further eastward, it was suppressed under the Cenozoic by the Xialiao River trough. Geophysical data indicate that it has good continuity. Some traces are evident in the Tieling region. Further eastward, it passes east of Liaoning, south of Jilin and north of North Korea, and enters the Sea of Japan. The deep-sea trough in the sea of Japan forms its eastern section. Traces can be found in the areas north of Hokkaido and the Honshu Islands of Japan.[4]

A characteristic of this system is that basement rocks are extensively exposed. The rock blocks and rock fragments composed of the Archaeozoic–Palaeoproterozoic high-grade metamorphic rock system are intermittently distributed along this zone. The corresponding medium-acidic rocks and rock and migmatite zones are well exposed. The ancient crystallization basement of the Yinshan–Tianshan E-W-trending tectonic zone was formed by a combination of them. The Meso-Neoproterozoic rock zones are also well exposed. However, in terms of the characteristics of lithofacies formation, deformation and metamorphism, there is strong deformation and metamorphism in the Meso-Neoproterozoic fold zone in the Tianshan–Beishan section (possibly including the Alashan block). These features constitute the regional fold basements formed by the Jinning Movement. The Yinshan–Yanshan section has a different development history. This zone, together with the region to its south, is the North China block. It was a stable Meso-Neoproterozoic caprock deposit without obvious tectonic deformation, metamorphism or magmatic activity before the Indosinian movement. In addition, it was in a relatively stable depositional and uplift environment for a long time. After the late Hercynian–Indosinian movement, the western, eastern and central sections gradually formed a unified Yinshan–Tianshan E-W-trending tectonic system, which was enhanced and developed by the subsequent Yanshan Movement and the Himalayan Movement. Since this tectonic zone experienced many tectonic movements, contained some old rock blocks and slices, merged some early tectonic traces and forms, and underwent extensive recombination with and reformation by the other tectonic systems in the later stage, it became complicated with diverse deformations. The ductile–brittle deformation development and the low-temperature high-pressure dynamic deformation and metamorphic zones are large in scale and well preserved.

Hence, only a few representative sections are selected for a brief introduction.

(1) Tianshan Tectonic Zone

The spatial distribution of the Tianshan–Yinshan E-W-trending tectonic zone (or the Tianshan E-W-trending tectonic zone) in Xinjiang moved approximately 100 km to the north. This zone is located at approximately 40°N–44°N, and some sections are located further north. The zone is located between the Junggar block and the Tarim block, with a nearly E-W-trending distribution. In the west, it passes through Kazakhstan and Kyrgyzstan and extends further west. In the east, it passes through the Xinjiang and Beishan Mountains in Gansu and is covered by the Badain Jaran Desert. The Tianshan E-W-trending tectonic zone has a complicated structure. In addition to containing the E-W-trending main tectonic traces and shapes, it intersects obliquely with two groups of the NW-NWW-trending dextral torsional and torsional compressive fault zone and the NE-NEE-trending sinistral torsional and torsional compressive fault zone, which are relatively well developed and large in scale. They developed mainly from two sets of torsional fault planes in the E-W-trending compressional tectonic zone, extending in N-S-W and N-E-E strikes in the early period and in N-E-E and N-W strikes in the later period. In addition, there is also a N-W-trending tectonic system compounded with it. This zone merged and contained some E-W-trending tectonic traces and forms (intermittently exposed along the axis of the compound anticline zone) generated in the pre-Paleozoic and currently distributed in the E-W-trending and some E-W-trending deep lithospheric fault zones. Additionally, there are some nearly E-W-trending Mesozoic and Cenozoic basins and troughs intermittently distributed along the zone, forming an E-W-trending compound syncline zone. Based on its development history and distribution features, the Tianshan tectonic zone consists of the following secondary zones in order from north to south: the Alatao–Bogda–Haerlik fold-fault zone, the Gongnaisi–Xinyuan depression fold zone, the Turfan–Hami intermountain depression zone, the Harkeshan–Bayinbuluke fold-fault zone, the Jueluotag–Heiyingshan fold-fault zone, the Kuluktag–Mazongshan uplift zone and so on.

The Kuluktag–Mazongshan uplift zone is located on the northern margin of the Tarim Basin and is a fault-fold uplift zone on the southern margin of the E-W-trending tectonic zone of the Tianshan Mountains. The surface section starts from Korla in the west, passes through Kuruktag and Xingxing Gorge, extends east of Wudaoming in Mazong Mountain of Gansu, and spreads under the Badain Jaran Desert. Further eastward, it connects with the Yinshan uplift zone in Inner Mongolia. The oldest crystalline basement rock exposed in the region is the Tuoge complex. Its lower section comprises a set of mid to high-grade metamorphic amphibolite facies, and the upper section is unconformably covered by the Paleoproterozoic Xingditag Group. The protoliths of the Xingditag Group are medium-basic volcanic and terrigenous clastic rocks.[5] The folds, schistosity and gneissosity of the Archean–Palaeoproterozoic are distributed mostly in nearly E-W-trending rocks. Both the Qingbaikouan and Sinian systems of the Neoproterozoic are exposed in this region. The Palgangtag Group of the Qingbaikouan system is composed of stable neritic platform facies sandstones and dolomite limestones with stromatolite. They are well exposed in the Beishan Mountain region.

Neritic carbonate, carbonaceous and argillaceous deposits were widespread in the early Paleozoic, with high contents of phosphorus, uranium and vanadium at the bottom. Afterward, the middle Ordovician South of Kuruktag–Fangshankou region was composed of mega-formation turbidites. The lower part of both the northern and southern sides of the Mazong Mountain compound anticline is dominated by graptolite facies, and the upper part is dominated by fossils of cephalopods and trilobites, both of which belong to the South China biota. In the Silurian, there was an active graptolite-bearing shale formation that transitioned upward to land–sea alternate facies and red continental clastic rock deposits with a transitional relationship with the middle-lower Devonian molasse formation. At the end of the Devonian, this region folded and consolidated to form a relatively stable continental crust. From the late Paleozoic to Mesozoic and Cenozoic, it was uplifted for a long time and subsided only locally in the Kuqa and Luntai regions, forming tension-fault basin sedimentation. However, the Carboniferous in the Beishan region was an active marine deposit dominated by clastic rocks, limestones and volcanic rocks, with thicknesses of more than 6,000 m, indicating that a unified E-W-trending deposition environment did not form in the Carboniferous. The early Permian comprised marine medium-basic, acidic and basic volcanic rocks. Late Permian terrestrial volcanic rocks were mainly medium acidic, followed by medium basic rocks. They developed mainly in the southern part of Beishan Mountain and became distributed in the Liuyuan–Dacang fold zone, forming a volcanic rock zone as long as 100 km. The intrusive rocks in this zone were well developed, and extrabasic, basic, medium acidic and alkaline rocks were all exposed, but the granite was dominant.

(2) Yinshan–Yanshan Tectonic Zone
The middle section of the Tianshan–Yinshan E-W-trending tectonic zone extends eastward from Alashan, passes through Langshan and spreads along the Yinshan Mountains and Yanshan Mountains to the west of the Xialiaohe trough, with clear E-W-trending tectonic traces and

good continuity. It is a part of the main bodies of the Tianshan–Yinshan E-W-trending tectonic system. Its northern boundary extends mostly along the border between China and Mongolia and passes through Suolun Mountain, Erdaojing, Chagan Nuoer and Darinuoer to the Songliao Plain along the Xilamulun River. Its southern boundary extends from the north side of Yabulai Mountain in the Badain Jaran Desert, passes through Dengkou, the north side of Dongsheng Uplift and Taihang Mountain, to the north margin of the North China Plain and then enters Bohai Bay. It was cut and reformed by the N-N-E-trending Helanshan–Jinpingshan fault uplift zone and the Xing'an–Xuefeng fault uplift zone somewhere on the route, with some sections appearing scattered and intermittent. However, it is generally a large-scale E-W-trending tectonic deformation and metamorphic zone and magmatic activity zone with good continuity. Based on the tectonic development history and the features, it is divided into northern and southern subzones.

The northern subzone lies between 42°00'N and 43°40'N and spreads into the area spanning Solun Mountain, Modula, Wendul Temple, Wengniute Banner, Kure Banner and so on, measuring 1,320 km long from east to west and 50–200 km wide from north to south. Its northern boundary extends westward from Erlianhot and the southern part of Sulente Left Banner to the Xilamulun River. Its southern boundary extends from Langshan to Agui, Huade and Chifeng north of Baiyun Obo and then to Zhangwu, with a wave-like shape in the E-W extension, which is known as the trough and platform boundary. It is also a boundary fault zone and a lithofacies abrupt zone. In order from south to north, the main constituents of the northern tectonic subzones include the Xianghuang Banner–Kure Banner fold-fault zone, Suolunshan–Linxi fold-fault zone, Ailige Temple–Erdaojing fold-fault zone, Xilamulun River fold-fault zone and thrust nappe tectonic zone in the middle of Sunite Left Banner. Among them, the Sangendalai Cretaceous faulted basin and the Otindag Cenozoic rifted trough developed. The long axes of the secondary uplifts and depressions in the trough also show E-W-trending, which are often controlled by the E-W-trending hidden faults. This subzone has the following deformation features: the fold zone is the main body; the nappe tectonics east of the Yinshan Mountains developed mainly on the northern and southern sides of the Suletu Jurassic coal-bearing basin in Chahar Right Middle Banner. The Heiniugou–Pangyangshan–Ulanheya thrust zone is located on the north side, with a length of more than 50 km in the E-W extension. The fault section tilts northward, and the old strata have

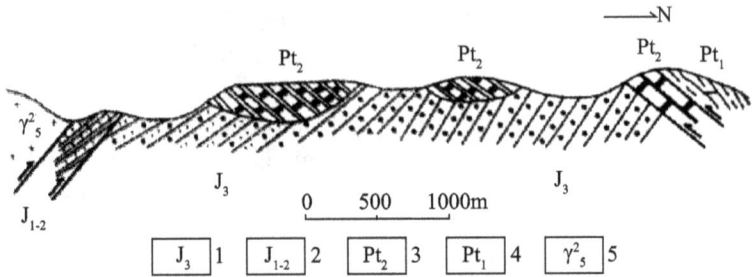

Fig. 2.3.   Cross-sectional diagram of the Xiaobaituozigou thrust nappe tectonics in the Suletu region.

been layered over by new strata from north to south, forming the klippes. The thrust zone on the south margin of the Suletu Basin extends more than 60 km from east to west, and all sections tilt southward. It forms N-S-trending thrust nappe tectonics with the thrust fault in the north (Fig. 2.3).

As mentioned above, in the uplift zones in the Yinshan Mountain region, the thrust nappe tectonics on the northern and southern sides of the upper Carboniferous-middle-lower Jurassic coal-bearing basin were well developed. All of them are imbricate thrust nappe tectonics. This feature indicates that multiple strong N-S-trending horizontal compressional movements of the vertical mountain chain have occurred in this region since the Mesozoic.

In addition, the Kober tectonics may exist in the middle of the orogenic zone. For example, in the south of Sanheming, some regions of the two groups of E-W-trending fault zones are unconformably covered by the Cretaceous basin. The northern side of the fault plane tilts southward, and the Erdaowa Group overlies the Hercynian granite from south to north. The south fault plane tilts northward, and the Paleoproterozoic Erdaowa Group overlays the Mesoproterozoic Baiyun Obo Group from north to south, forming a reverse thrust zone with an intermittent extension of 180 km. To the west, kyanite schist appears in the Shujigou Formation of Tanertai Mountain. Along the Longhua–Damiao faults in the north uplift zone of the east wing, there are also kyanite mylonitic quartz schists, migmatization formations (isotopic ages of 230 Ma and 236 Ma) and alkaline granites (217 Ma and 223 Ma).

In summary, it is inferred that the central axis of the Indosinian orogenic zone might be located at approximately 41°00'N. The finalization period of the Yinshan–Yanshan E-W-trending tectonic zone was neither the Precambrian nor Yanshanian, but the main episode of Indosinian movement in the Indosinian Epoch.

(3) Liaodong–Jidong Tectonic Zone

This section refers to the region east of the Xialiao River and is the eastern extension of the Tianshan–Yinshan E-W-trending tectonic zone. Its northern boundary is approximately along the line of Changtu–Panshi and Huadian–Antu–Wangqing at approximately 43°00'N. Its south boundary is along the line of Wafangdian–Zhuanghe in the south of Liaoning at approximately 40°00'N or further south. Since it was once cut and reformed by the NE-NNE-trending tectonic zone, the E-W-trending tectonic traces are scattered, and the locations have also changed to varying degrees.

The Shuangyang–Yanbian fold-fault zone of the northern subzone includes the (1) Kaiyuan–Meihekou fault zone, (2) Shuangyang–Yanbian fault fold zone and (3) Antuxinhe–Hunchun Madida fault zone.

The Tieling–Taizihe fold-fault zone of the south subzone spreads within the North China block. Its main constituents include the (1) Tieling fold zone; (2) Taizihe depression-folding zone; (3) Xinghua–Baitoushan Tianchi fault zone; (4) Daquanyuan–Changbaishan fault zone; (5) Nangushan Yanshanian complex zone; (6) Berlinchuan Indosinian alkaline complex zone (208–223 Ma); (7) Shuangyashan–Dabao Indosinian magmatic rock zone (220–226 Ma); (8) Buyun Mountain fold zone; (9) Furongshan tectonic magmatic rock zone (207 Ma); (10) Wafangdian–Zhuanghe tectonic magmatic rock zone (200–213 Ma); and (11) Jinzhou–Dongjiagou ductile fault zone.[6]

(4) Tianshan–Yinshan E-W-Trending Tectonic Zone

Many studies have shown that the Tianshan–Yinshan E-W-trending tectonic zone has not existed since the Archean. It was formed and finalized in the late Hercynian-Indosinian period after experiencing many tectonic movements. It was enhanced in the Yanshanian period, accompanied by multistage sedimentation, metamorphism, magmatic activity, crust-mantle material evolution and mineralization and orogeneses with alternating compression and extension. Its evolutionary history in various stages was not the same. With Yinshan–Tianshan Mountain as the central axis, it formed and developed from south to north and from west to east. The Tianshan–Beishan section was finalized by deformation and metamorphism in the late Hercynian period, but its features were not obvious in the Indosinian. There was an Indosinian medium acidic rock intrusion in the Alashan region, indicating the existence of the Indosinian movement. To the east of Langshan, the Indosinian movement was the main deformation period. However, the prototypes appeared in the early-Mid-Hercynian period and constituted the main boundary between the Beihai south land or the northern active zone and the southern stable zone in the Carboniferous-Permian.

In terms of the relationship between formation and tectonics, formation is controlled by tectonics, and tectonics is reflected in formation to a certain extent. That is, a giant deformation zone is the precondition of a deposition formation, and the lithofacies, thickness and characteristics of the formation in the formation process reflect the characteristics of the distribution and evolution of the giant deformation zone to a certain extent. The restoration of ancient positions in different periods needs to be further studied. Only the current distribution is discussed briefly here.

After experiencing the Luliang Movement, the North China block exhibited discontinuous E-W-trending uplifts and depressions in the middle Neoproterozoic epoch. The middle and eastern sections were superimposed by the N-E-trending uplift and depression zones, and the western section was superimposed by the N-W-trending uplift[7] and depression zones. Different types of deposits formed in the Mesoproterozoic–Early Paleozoic.

In the south subzone, the main orogenic movement occurred in the early Indosinian, which not only involved the Middle Neoproterozoic–Middle Triassic strata in folds but also involved the ancient crystalline basement blocks and rock blocks to a certain extent. The Yinshan region caprock folds and basement folds were E-W-trending tectonics, which were mutually reconnected. This set of folds experienced low-grade greenschist facies low-temperature dynamic metamorphism. Glaucophane, kyanite, hard chlorite, etc., appeared in the E-W-trending deformation and metamorphic zone, which was the medium- and high-pressure dynamic metamorphism zone. In addition, the ductile deformation zones in Xidoupu, south of Sanheming and Jinzhou in south Liaoning indicated that the Indosinian deformation was middle-deep, as deep as reaching the lower crust and upper mantle. In the Tianshan section, there is a ductile shear zone of more than 600 km along the Qiugemingtashi–Huangshan line, which developed in the Upper Paleozoic. Exposure of glaucophane schists has been found along the Bayanbulak zone and other places. According to the results of the finite strain measurements in the Yanshan region, the *z*-axes of the Mesoproterozoic Wumishan Formation and Hongshuizhuang Formation decreased by 17–24%. The *z*-axis of the middle Carboniferous-Lower Triassic decreased by 47–57%. The value of the regional paleodifferential stress was 20–50 MPa.

In summary, in the formation and evolution process, the Tianshan–Yinshan E-W-trending tectonic system experienced the alternation of N-S-trending tension and compression systems many times. Its formation period (or finalization period) was the Late Hercynian–Early Indosinian. A composite orogenic zone was formed by merging, inclusion, collage and welding. The main orogenies were rift-closing orogeny in the deposition period, compression orogeny, ramp thrust and Kober orogeny, and thermal uplift orogeny in the finalization period. During the late Indosinian–Yanshanian period, an E-W-trending inland coal-bearing and petroleum-bearing basin formed along the middle section of the Tianshan–Yinshan E-W-trending zone, with relatively large folds and thrust nappes.

2.2.3   *Kunlun–Qinling E-W-Trending Tectonic System*

The Kunlun–Qinling E-W-trending tectonic system spreads across the central mainland of China, starting from Pamir in the west and reaching the south of the Yellow Sea in the east, with clear traces, good continuity and large scale. The total length is more than 4,000 km. It has been the natural boundary of the north and south geological tectonics of Mainland China since the Late Paleozoic. It also represents the common development of the North China block and the Tarim–Qaidam block. In the Late Paleozoic and Early Mesozoic, the axis of the zone was the main boundary between the South Sea and northern land. Due to the superimposition, reformation, composite formations and cutting by the later tectonic systems and the heterogeneity of block displacement of all the sections, its tectonic landforms are complicated, with undulating strikes and greatly different latitude locations. Due to the uplift of the Pamirs Plateau and the formation of the Hetian arc tectonics in south Xinjiang, the western section moved significantly northward to approximately 37°N. The east Kunlun and Qinling sections spread at 32°N–35°N. This tectonic system extends eastward, passes through the South Yellow Sea and Jeju Strait and connects with the Honshu E-W-trending tectonic zone in Japan. In the west, this tectonic system joins with the Kabul–Tehran E-W-trending tectonic zone after passing through the central Pamir Plateau. Farther west, it might connect with the Mediterranean E-W-trending tectonic zone. This zone is strongly deformed and metamorphic. The thrust nappe tectonics are well developed. It basically reflects the deformation features of the Kunlun section.

In terms of this tectonic system within Mainland China, the "Kun-qin Zone" can be approximately divided into the following sections, which will be introduced briefly in order from west to east in the following paragraphs.

(1) West Kunlun Mountain Tectonic Zone

This section refers to the western Kunlun Mountain region west of the Altyn–Kuyake fault zone. It is a giant reverse S-shaped fold mountain system and fault uplift zone at the southwest margin of Tarim Basin. In the south, it is adjacent to the Kangxiwa deep fault and Karakoram fold system. In the north, it is adjacent to the southwest depression of Tarim Basin.

Its main body spreads at 35°N–37°N. This region has the same crystalline basement and fold basement as the Tarim Basin, belonging to the continental crust of the Tarim block. It can be divided into the following three zones in order from north to south.

(a) *Tiekrik fault uplift zone*: The Tiekrik fault uplift zone, or the North Kunlun fault uplift zone, is located on the southwest margin of the Tarim Basin. It is an arc-shaped zone protruding southward, adjacent to the Selagaz north fault, the Kegang fault, the depression on the south margin of Tarim basin and the Kallong and Kurlang fold zones in the south and north, respectively. It is a stable block embedded between both zones. Presinian formation constitutes the main body of the fault uplift zones, and Sinian and Paleozoic formations are stable caprock deposits distributed along the margin of the fault uplift zone. The Palaeoproterozoic formation is a middle- and high-grade metamorphic rock system dominated by schists and gneisses. The Meso-Neoproterozoic formation has relatively low-grade metamorphism. The Sinian formation is unmetamorphic dolomitic and siliceous carbonate and clastic rocks. The Middle-Upper Ordovician and Upper Paleozoic formations are neritic-sea–land alternation facies clastic and carbonate rocks. There was weak magmatic activity in this zone, and only small rock strains, such as diorite, granodiorite and monzonitic granite, were found in a near E-W-trending zone in the

Paleoproterozoic, covered unconformably by Carboniferous sediments, which might be the products of the Caledonian–Hercynian period.

The tectonics of the fault uplift zone are complicated. The pre-Sinian folds are tightly closed. Its axial direction is basically the same as that of the mountain. The west section has a N-W-W strike, and the east section has a nearly E-W strike. Both have overturning tendencies toward the basin. The Sinian and later strata have wide and gentle folds. The development of faults can be divided into two groups. One group is distributed in a nearly E-W-trending direction, which is basically consistent with the axial direction of the folds. Most of faults in this group are compressional thrust faults toward the basin and formed earlier. The other group comprises N-N-W-trending faults, characterized by dextral strike-slip and formed in the Mesozoic and Cenozoic. The former constitutes the Kunlun E-W-trending tectonic zone, while the latter might be the peripheral component of the head of the Pamir-Himalayan reverse S-shaped tectonic system. They were created mainly by the merger and reformation of the Hetian arc-shaped tectonic zone and play an important controlling role for the depression basin on the southwest margin of Tarim.[8] The early history of this zone could be the Mesoproterozoic intercontinental activity zone. In the east, it is connected with the Altun Mountain intercontinental activity zone, which was once an oceanic crust activity zone halfway around the Tarim block. After experiencing folding and uplift by the Altun–Tarim movement, it became an arc-shaped tectonic zone in the southern part of the Tarim block, forming the base of the Hetian arc-shaped tectonic zone. In the Paleozoic, especially in the late Paleozoic, the front section of the arc-shaped uplift fault zone merged and reformed as a constituent of the Kunlun E-W-trending tectonic system, forming the Tiekrik E-W-trending fault uplift zone.

(b) *Challon–Kurlang fold zone*: In the north, this zone is adjacent to the Kegang fault zone and the Tiekric fault uplift zone. In the south, it is adjacent to the Bulunkou fault zone, the Karak fault zone and the Gongger–Sangzhutag uplift zone. It extends westward from the border after passing through Pamir. In the east, it disappears near the Buya area. It is a rift-type active fold zone formed after the disintegration of the Tarim block. Its basement is in the continental crust of the Tarim block. At the end of the Ordovician, the submarine grabens formed between the north and south faults. In the middle and late periods of the early Permian, the folds formed after many Silurian–Carboniferous movements. The basement of the folds is composed of Mesoproterozoic and Neoproterozoic formations. Most of the rocks are medium-pressure amphibolite-amphibolite-facies metamorphic rocks. The middle-lower Ordovician formation is a stable deposit. The Silurian and middle Devonian rocks have the characteristics of flysch-like formations. The Upper Devonian formation comprises sea–land alternation facies and a molasse-like formation, which is integrated with and transitional to the lower Carboniferous formation. The latter consists of carbonate rock formations, clastic rock formations and medium-basic volcanic rock formations. A set of deep-sea pillow basalts (2,100 m thick), iron siliceous rocks, andesites, dacites, felsites and tuffs developed along the Gulgunek ditch. The middle-upper Carboniferous is composed of littoral-neritic clastic rocks and carbonate rocks, which are rich in fossils. The lower Permian formation is a molasse that is unconformably integrated over the Carboniferous formation. The Upper Triassic formation is composed of continental coal-bearing formations and continental volcanic formations.

Magmatic activities in the region are mainly from basic volcanic eruptions and acidic magma intrusions in the Carboniferous and Permian. Mega-thick pillow-shaped basalt rocks and ferro-siliceous rocks appeared in the Carboniferous. The volcanic activity followed the evolutionary law of continental crust-transitional crust-oceanic crust. The middle Hercynian intrusive rocks are mainly calc-alkali series orogenic granitoids, with isotopic ages between 297 and 364 Ma. The late Hercynian intrusive rocks are plagioclase granitoids, which are calc-alkaline, weakly saturated aluminum and have the characteristics of multiphase differentiation. They might represent the Permian extensional environment and be the products of inland rifting.

The main deformations in this subzone include the Aksayibashi Mountain complex syncline zone, Qallong complex anticline zone, Kurlang complex anticline zone, Kiziltao depression zone and some main fault zones among them. The complex folds formed in the late Paleozoic. The intermountain depression zone was also the Indosinian depression based on the Hercynian folds.

(c) *Gongger–Sangzhutag uplift fold zone*: This uplift fold zone is located in the Gongger–Mostag–Sangzhutag region. In the north, it is adjacent to the Bulunkou fault zone, Karak fault zone and Charlong–Kurlang fold zone. In the south, it is bounded by the Kangxiwa super-lithospheric fault, which is separated from the Karakoram fold system. In the east, it reaches the western side of the Kuyake fault (the southern section of the Altun fault zone). In the west, it extends west of Pamir.

This uplift fold zone is composed of mostly a Precambrian system, with only a few Lower Paleozoic caprocks. The Middle Ordovician and Silurian subactive marine metamorphic clastic rocks intercalated with carbonate rocks are unconformably on the Great Wall system.[9] The upper Devonian is a set of continental-sead/land alternation facies red fine clastic rock formations.

This uplift zone used to be an integral part of the Tarim block in the Precambrian. The basement fault zone along the northern and southern sides of the uplift fold zone subsided by extension in the Caledonian period, and a graben formed in the Challon–Kurlang region. The zone further developed and expanded in the Hercynian period, separating from the Tarim block. It remained in an uplifted state for a long time and lacks Carboniferous, Permian and Mesozoic deposits. In the late Permian, this zone was once again integrated with the Tarim block. During the Himalayan movement, due to the impact of the Pamir-Himalayan reverse S-shaped tectonic system, basic volcanic eruptions in the Cenozoic developed along deep faults; in particular, the boundary fault-Kangxiwa super-lithospheric fault zone became more prominent. Therefore, in addition to the second-stage Proterozoic gneissic granites covered unconformably by Qingbaikouan in the middle of the uplift fold zone, this uplift fold zone is composed of mainly the Hercynian-Indosinian intrusive rocks developed along the northern side of the Kangxiwa fault. The isotopic ages of the medium-acidic rocks are 202. 3–277.7 Ma and 297–346 Ma. The well-developed ophiolitic mélanges are distributed along the Kangxiwa super-lithospheric fault zone in the Kudi region. This uplift fold zone was previously known as the "West Kunlun uplift zone".

(2) East Kunlun Mountain Tectonic Zone

This section includes the entire Kunlun Mountain system to the east of the Kuyake fault zone. Its main body is distributed along the southern margin of Qimantag, the southern margin of Qaidam Basin and in the Hoh Xil Mountains. In the east, it includes the nearly E-W-trending tectonic traces and shapes in Buqing Mountain and Xiqing Mountain. As this region was restricted by the N-W-trending tectonic system in the early period, superimposed and reformed by the Qinghai–Tibet–Sichuan–Yunnan reverse

S-shaped tectonic system, and interfered with by the Sichuan–Yunnan N-S-trending tectonic system in the later period, its tectonic traces and deformation and metamorphism are complicated and discontinuous. One part of the zone is obliquely reconnected and combined with the N-W-trending tectonic zones, such as the Qimantag fold zone, the Bokarek–Animaqing fault zone and the Animaqing fault fold zone. The other part is obliquely reconnected and combined with the NWW-EW-trending tectonic zones of the Qinghai–Tibet–Sichuan–Yunnan reverse S-shaped tectonic system, such as in the Kunlun Mountains and the Hoh Xil Mountains and the regions as far as north of the Qumalai Bayankla Mountains. In the Ruoergai–Xiqing Mountain region, the E-W-trending tectonic zone is still disturbed and affected by the Sichuan–Yunnan N-S-trending tectonics, so its continuity in this region is poor but still continues intermittently. The main E-W-trending tectonic traces and forms in this section include the Qimantag–North Kunlun tectonic zone, middle Kunlun main ridge fault zone, Alkashan–Burhanbuda fold zone, Muztag–Jingyuehu–Xiugou fault zone, Hoh Xil complex fold zone and Xiqing Mountain–Jishi Mountain fault fold zone.

(3) West Qinling Tectonic Zone

In the western Qinling region, the Qinling–Kunlun E-W-trending tectonic system is distributed mainly south Gansu, i.e., between 32°30'N and 35°N. A part of it is distributed northwest Sichuan in an arc form protruding southward, with latitude slightly south. On the basis of the characteristics of the deposition formation, magmatic activity, metamorphism and tectonic deformation, it can be approximately divided into northern–southern subzones along the Weihe Depression zone.[10]

The rock formation of the northern subzone consists of mainly the Niutouhe Group of the pre-Great Wall system and Great Wall system. Except for the Devonian system, it can be seen sporadically in other Paleozoic strata. The southeast section of Tianshui south of the Gangu–Wushan fault zone is located in the northern Qilian fold zone of the early N-W-trending tectonic system and near the arc top of the western wing of the Qilu–Helan epsilon-shaped tectonic zone. These zones are obliquely connected-reconnected and combined with E-W-trending tectonics, with indistinguishable tectonic traces. However, the related magmatic rock zones have their own spatial distribution. For example, the Caledonian medium acidic magmatic rocks obviously follow a N-W-W-trending zonal distribution, while the Hercynian, especially Indosinian and Yanshanian, medium acidic magmatic intrusions are in an E-W-trending zonal distribution. The Tongren–Wushan–Gangu fault zone is the boundary along which the Hercynian extrabasic rocks are distributed. After the Yanshan Movement, this region gradually became an inland depression, the Weihe Depression. Along the Weihe River, there are extensively developed mega-thickness Mesozoic and Cenozoic continental clastic formations and loess formations. Due to interference and cross-cutting by the early N-W-trending system, the later Qilu–Helan epsilon-shaped western wing, the late Longxi torsion tectonic system, the western margin of the N-N-E-trending tectonic system and the features of the north subzone are not conspicuous.

The southern subzone is located at 32°30'N–34°30'N, adjacent to the Weihe Depression in the north. In the south, it reaches the Hongyuan Arc northwest Sichuan. It measures more than 500 km long from east to west and 100–200 km wide from north to south. The eastern section was combined with the Qilu–Helan epsilon-shaped front arc fold zone and the N-N-E-trending system. The western section was combined with the N-W-trending tectonic system, resulting in the deflection of the tectonic line in the Xiqing Mountain region from E-W-trending to N-W-trending from east to west. These portions obliquely connected-reconnected and combined in the Jishi Mountain region, which is reflected in the deposition formation. In addition, the Xiqing Mountain region was also affected by the early N-S-trending zone and affected by the activities of the late Cenozoic N-S-trending zone, which is indicated by the N-S-trending seismic activity zone. This subzone is generally E-W-trending but sometimes spreads in an arc shape protruding southward. It is the main body of the Qinling–Kunlun E-W-trending zone in this section. The subzone is dominated by folds, faults and Hercynian, Indosinian and Yanshanian rocks composed of Paleozoic and Triassic marine strata. Based on the features of the tectonic traces and forms, it is divided into four tectonic zones in the Gansu Regional Geological Records in order from north to south: the Meiwuxin Temple–Dacaotan complex anticline zone, Luqu fault zone, Maqu–Wudu fault zone and Wen County–Kang County fold-fault zone.

In summary, the E-W-trending tectonic system in the western Qinling region occurred in the late Paleozoic, matured and was finalized in the Triassic. Geological activity continued after the Triassic but was relatively weak.

(4) East Qinling Tectonic Zone
The E-W-trending tectonic zone in the East Qinling region spreads between the North China block and the Yangtze block at 31°50'N–34°40'N. The northern basement is the North China type, and the southern basement is the Yangtze type. This zone was obliquely connected and combined on the Qilian–Qinling–Tongbai N-W-W-trending paleotectonic zone. The northern section was obliquely connected-reconnected and compounded with the Qilu–Helan epsilon-shaped front arc tectonic zone and reversely connected and compounded with the N-N-E-trending and N-S-trending tectonic systems. The western section was also compounded with the Longmenshan N-E-trending system, and the southern section was affected by the Dabashan arc tectonic system. Therefore, this zone's tectonic features are relatively complicated. This zone not only contains some ancient (pre-Paleozoic) land blocks and rock blocks but was also superimposed and reformed by later tectonics. Consequently, there are many differences in understanding regarding the tectonic framework and the evolutionary history of this region. According to the existing data, the Qinling fold system, as it is commonly called, should actually include two major tectonic systems. One system is the currently N-W-W-trending Caledonian-Meso-Neoproterozoic tectonic zone or Qinling–Qilian–Barokonu Caledonian trough fold zone. It is classified in the main tectonic zones of the N-W-trending system in this book, which will be discussed while reviewing the N-W-trending tectonic system. The other system is the Hercynian–Indosinian Qinling–Kunlun E-W-trending tectonic system, which is the main topic discussed in this section. In the Yanshanian and Himalayan periods, the N-N-E-trending zone showed obvious inherited activities from the influence of the N-N-E-trending system in the eastern Pacific tectonic domain.

The East Qinling Mountains are mainly in the Shaanxi region. This zone extends eastward to Henan and submerges under the North China Plain. In the west, it connects to the western region of the Qinling Mountains in south Gansu and northwest Sichuan. It constitutes an important dividing line in the geology, geography and climate between the northern and southern regions of China. Based on the characteristics of the formation and evolution, deformation and metamorphism and magmatic activities of the geological tectonic system in this region, the E-W-trending tectonics in the East Qinling Mountains are divided into the following four fault zones in order from north to south: Baoji–Hua County, Zhifang–Yongfeng, Taibai–Gao County and Zhitian–Shanyang.

2.2.4 *Nanling E-W-Trending Tectonic System*

This system is roughly located at 22°30'N–26°30'N and 99°00'E–120°00'E. Its main body is distributed at 23°30'N–25°30'N. It starts from the Hengduan Mountains on the China–Myanmar border in the west, passes through the Nanling Mountains and 7 provinces in southern China and extends eastward to Chinese Taiwan. It is more than 2,000 km long and approximately 200–300 km wide. Due to severe interference by other tectonic systems, the tectonic traces of this tectonic system are relatively broad, scattered and intermittent. The indications of the traces are not as obvious, prominent and continuous as in the Yinshan–Tianshan and Qinling–Kunlun E-W-trending tectonic systems. This tectonic system is composed of mainly E-W-trending folds, thrust fault zones, granite zones and metamorphic rock zones. Its overall appearance indicates that it is a compound uplift fold zone. Indications in its western section are not obvious due to strong interference by the Sichuan–Yunnan N-S-trending tectonic system, while indications in its eastern section are quite prominent. It geomorphologically constitutes a natural watershed between the Yangtze River and the Pearl River systems. The Nanling E-W-trending tectonic system formed mainly in the Mesozoic, and it is a relatively young giant E-W-trending tectonic zone. Based on its distribution features, it is generally divided into eastern, middle and western sections.[11]

(1) Western Section — Yunnan–Guizhou–Guangxi Section

The western section is located in approximately the Yunnan, western Guizhou and western Guangxi regions between the West Yunnan (Sanjiang) and Sichuan–Guizhou N-S-trending tectonic systems at 23°N–26°30'N. It is composed of mainly nearly E-W-trending compound uplift fold zones and thrust fault zones. Due to strong interference from other tectonic systems, most of the western section is hidden. There are E-W-trending tectonic traces scattered in central and western Yunnan between the West Yunnan (Sanjiang) and Sichuan–Yunnan N-S-trending tectonic systems. The gravity data indicate that the south boundary of this E-W-trending tectonic zone is near 23°N, corresponding to the E-W-trending Moho anomaly gradient zone. The crust is thicker to the north of 23°N, indicating an E-W-trending tectonic zone in the deep location. In the Yunnan–Guizhou–Guangxi region between the Sichuan–Yunnan and Sichuan–Guizhou N-S-trending tectonic systems, the E-W-trending tectonic traces are more obvious, and the compound fold zones and thrust zones are relatively developed.

(2) Middle Section — Nanling (Hunan–Guangxi–Guangdong–Jiangxi) Section

The western boundary of this section is the Sichuan–Guizhou N-S-trending tectonic system. The eastern boundary is the Taining–Longyan N-S-trending tectonic zone of the Jiangle–Longyan N-S-trending tectonic zone, located in the Hunan–Guangxi–Guangxi–Jiangxi region at approximately 22°40'N–26°30'N and 108°E–117°E, with the main body located in the Nanling region at 23°N–26°N. This section is the main part of the E-W-trending Nanling tectonic system. It is composed mainly of E-W-trending compound folds, thrust fault zones and granite zones, as well as dynamic metamorphic zones, late Cenozoic uplift depression zones and N-S-trending tectonic refolding zones and other formations. This section has a large scale, clear traces and good continuity, with the magnificent E-W-trending granite zone as its prominent feature. It can be divided into the following five secondary tectonic zones (subzones) from north to south: Yangming Mountain–Shicheng, Jiuyi Mountain–Dayu, Yishan–Xunwu, Dayaoshan–Fogang–Fengshun and West Daming Mountain–Gaoyao–Huilai. The last three tectonic zones are the main ones, and the main zone is the Dayaoshan–Fogang–Fengshun E-W-trending tectonic zone.

(3) Eastern Section — Fujian–Taiwan Section

The eastern section is located in Fujian, Penghu Islands and the north-central part of Taiwan at approximately 23°30'N–26°30'N and east of 117°E. It is composed mainly of a series of E-W-trending thrust faults and some compound folds, granite rocks and volcanic basins. It also includes dynamic metamorphic zones, hot springs, epicenters, aeromagnetic anomalies and the distribution of some E-W-trending mountains and water systems. These forms are intermittently connected due to interference by other tectonic systems. Overall, this section can be divided into the following three E-W-trending secondary tectonic zones from north to south: Mingxi–Minqing, Zhangping–Xianyou and Nanjing–Xiamen, which approximately correspond to the E-W-trending tectonic zones of the middle section of the Nanling E-W-trending tectonic system. In addition, there is a Penghu–Taichung E-W-trending tectonic zone in the southeastern sea region.

The regional E-W-trending tectonic zones refer mainly to scattered, intermittent but slightly E-W-trending tectonic zones intermittently distributed among the three main E-W-trending zones in China. Among them, the relatively notable zones include the E-W-trending fold zones passing through the central part of the Ordos Basin, the central part of the Shanxi land block, the northern margin of Jiaodong Peninsula and the E-W-trending fold zone passing through Zhongwunong Mountain in northern Qinghai. They are located between the Yinshan–Tianshan and Qinling–Kunlun E-W-trending tectonic zones, with an intermittent length of approximately 2,000 km. The secondary regional E-W-trending tectonic zones are located between the Qinling–Kunlun and the Nanling E-W-trending zones at approximately 28°N–29°30'N. They appear intermittently in Ebian, Junlian in the southern Sichuan–Xuyong and Yueyang–Jiujiang fold zones, and central Tibet at approximately 30°N, with a length up to approximately 3,000 km.[12] In addition, the E-W-trending folds and faults are intermittently exposed in the west to Xingkai Lake, the middle section of the Greater Xing'an Mountains and the northern margin of Junggar at 46°N–47°N. They are located between the Yilahuli E-W-trending zone and the Tianshan–Yinshan E-W-trending zone. The Pingxiang–Xinyi fault fold zone on the southern margin of Mainland China and the fault magma active zone in the Leizhou Peninsula also show intermittent extension from east to west. Together with the main E-W-trending zones, they reflect the distribution of different levels of E-W-trending tectonic zones, which generally have equidistant characteristics.

The three large E-W-trending tectonic land blocks are located among the five major E-W-trending tectonic systems in China, i.e., the northern-most Junggar block and Songliao block, the land blocks of the Tarim–Qaidam Basin, Alashan Basin, Yinshan Basin, Shanxi and North China Plains in the middle and the land blocks of North Tibet, Sichuan and Jianghan in the south. They are mainly the subsidence zones among the fold-fault zones of the E-W-trending tectonic systems that formed in the late Hercynian and early Yanshanian. Since the mid-late Yanshanian, the eastern section has been divided by the uplift and subsidence zones of the Neocathaysian direct-shear torsional tectonic system, showing lozenge-shaped composite basins or uplifted blocks, while the western section has remained relatively intact.

As mentioned earlier, there is no important genetic relationship between the North China block and the Tarim block in the early Paleozoic and in the long early geological period. Only in the late Carboniferous, with the formation and development of the extensive E-W-trending tectonic systems on both the northern and southern sides, was the North China Basin integrated with the Tarim–Qaidam–Alashan Basin and the Ordos Basin. The three E-W-trending subsidence zones have played an important role in controlling the land–sea patterns of the continental crust in China since the late Hercynian. This pattern was changed by the intense activities of the Peri-Pacific active zones and the Alps-Himalayan Meso-Netethys active zones.

2.2.5 Xisha E-W-Trending Tectonic Zone

This tectonic zone is located at 15°N–19°N: in the west, the Dongrek Mountain tectonic zone in central Laos and northern Cambodia. It extends eastward through the Manila Trench and the northern part of Luzon Island and connects with the Benham Seamount. Then, it extends further east through the Central Pacific Mountains and the Clarion faults to the pelagic syncline in Central America. The uplift in the Haitian Island region in the West Atlantic Ocean, Puerto Rican Trench, the Southeast Arabian fault, the Niger River in Africa and the tectonic zone of the Senegal River basins are integral parts of this S-W-trending tectonic zone.

2.3 E-W-Trending Tectonic Systems in the Northern Hemisphere

The E-W-trending tectonic zones in the Northern Hemisphere are generally well zoned, with magnificent scales and high regularity. They can be divided into the following nine E-W-trending tectonic zones from north to south (Fig. 2.4).

(1) *Tectonic zones at* 68°N–70°N: The E-W-trending fold zone is distributed in Jokuldakh in eastern Siberia (140°E–160°E), Brooks of Alaska, North America (140°W–170°W) and Spitsbergen Islands in the North Atlantic Ocean. It was active in both the Paleozoic and Mesozoic and formed a continental marginal fault zone in the Cenozoic.

(2) *Tectonic zones at* 58°N–62°N: There are Archean and Proterozoic E-W-trending fold zones in the Aldan–Angara region of northern Asia on which the Mesozoic and Cenozoic nearly E-W-trending basins are superimposed. These zones extend eastward to the northern margin of the Okhotsk Sea and westward to West Siberia. There are a series of

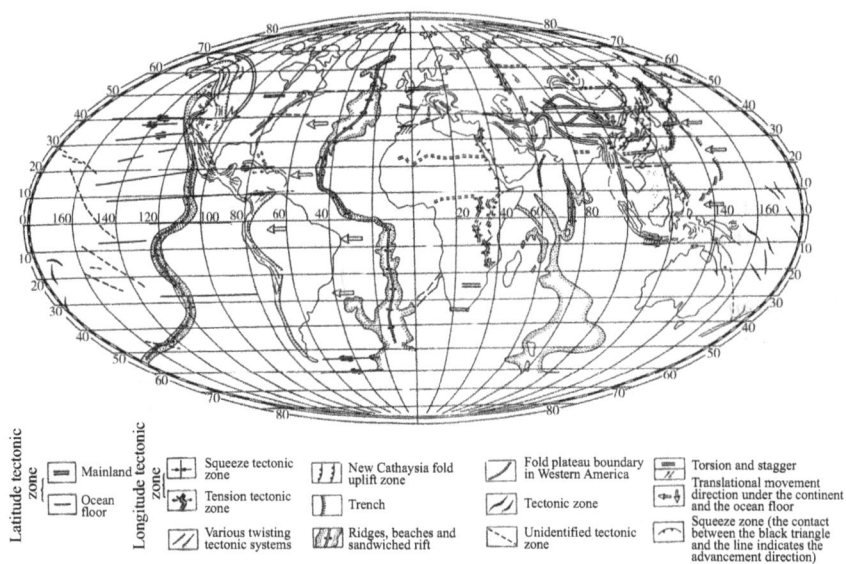

Fig. 2.4. Distribution of the main N-S-trending and E-W-trending tectonic systems on Earth (modified from Li Siguang, 1973).

E-W-trending rifts in the Finland Gulf at 20°E–30°E and the Hudson Bay in North America at 80°E–90°E, indicating strong activities since the Mesozoic.

(3) *Tectonic zone at* 50°N–52°N: Yilahuli or Tangnu–Kent tectonic zone, as mentioned before.

(4) *Tectonic zones at* 40°N–43°N: This zone refers to the Tianshan–Yinshan E-W-trending zone in China. It extends eastward to the northern part of Honshu Island in Japan. The Mendocino fault zone in the eastern Pacific, the E-W-trending gravity anomaly zone in the Cordillera Mountains in the western United States and the Einta Mountain uplift of the large anticline formed by the Precambrian system may be continuations of this E-W-trending zone in North America. The North Azores tectonic zone at the bottom of the Atlantic Ocean, the Pyrenees–Cantabria tectonic zone in southern Europe, the Anatolian zone near the Caspian Sea in northern Turkey and, further east, the intermittent fold zones in southern Fergana in central Asia are connected with the Tianshan fold zone in China.

(5) *Tectonic zones at* 32°N–36°N: This zone refers to the Kunlun–Qinling E-W-trending zone. It extends eastward to the Murray Fault zone in the east Pacific, the E-W-trending tectonic zone to the north of Los Angeles in the western United States, the Uchita fold zone in the southern United States (at 90°W–110°W) and westward to the E-W-trending Palo Pamisus Mountain fold zone from the Kunlun Mountains and through Afghanistan and the Atlas fold zone along the Mediterranean Sea in North Africa.[13]

(6) *Tectonic zone at* 23°N–26°N: This zone refers to the Nanling E-W-trending tectonic zone in China. It extends westward to the Ganges Fault (80°E–90°E) through the southern bank of the Brahmaputra River and the concealed E-W-trending tectonic zone in the Sahara Desert in North Africa. In the east, the Dianluogu E-W-trending submarine fault zone passes through the region near the Tropic of Cancer in the eastern Pacific. Further east, the E-W-trending fold zone in southern Pallas in central Mexico, the Caribbean Sea and the island of Cuba all belong to this tectonic system.

(7) *Tectonic zone at* 15°N–19°N: This zone refers to the Xisha E-W-trending tectonic zone in China. In the west is the Dongrek tectonic zone in central Laos and northern Cambodia. It extends eastward through the Manila Trench and the northern part of Luzon Island and connects with the Benham Seamount. Then, it extends further east through the Central Pacific Mountains and the Clarion faults, to the pelagic syncline in Central America. The uplift in the Haitian Island region in the western Atlantic Ocean, Puerto Rican Trench, the Southeast Arabian fault, the Niger River in Africa and the tectonic zone of the Senegal River basins are integral parts of this S-W-trending tectonic zone.

(8) *Tectonic zone at* 8°N–10°N: This zone is mainly a giant compression and torsional tectonic zone spreading along the southern coast of the Caribbean Sea and north of South America. It extends through the narrow continental ridge in Central America to the Pacific Ocean and connects to the giant Kripatong strike-slip fault zone.

(9) *Tectonic zones at the equator*: The Galapagos fault zone crosses the Pacific ridges, extends eastward through the northern part of South America and Romance in the Atlantic Ocean and connects to the Chayin compression and torsion fault zone. There is an E-W-trending fold-fault zone composed of the strata of the ancient Tertiary system and earlier in

the central part of Kalimantan. It is distributed along an E-W strike east of Sulawesi and the Sula Islands along the composition line of Seram Island. There is an E-W-trending sinistral Solang fault zone on the northern margin of West Irian near Sulawesi.[14]

2.4 E-W-Trending Tectonic Systems in the Southern Hemisphere

Due to the limited land area in the southern hemisphere, the E-W-trending tectonic traces are distributed at the bottom of the waters. This tectonic system can be approximately divided into the following zones.

(1) *Cerro De Bosco E-W-trending tectonic zones at 7°S–10°S in the Pacific Ocean*: The nearly E-W-trending tectonic traces are distributed west and east of the South American continent, the Atlantic ridges, Java Trench of South Asia and the Nugadenla Islands.

(2) *Tectonic zones at 17°S–20°S of the Arica Point tectonic system on the west coast of South America*: The South Rodriguez fault zone in the Indian Ocean ridge, etc.

(3) *Tectonic zones at 23°S–27°S south of Delanwa in South Africa*: The Musgrave tectonic zone in central Australia, the Easter fault zone in the eastern Pacific on the Tropic of Capricorn and the Petre Rios-Elivaldo copper mine zone in South America are also in this zone.[15]

(4) *Cape tectonic system in southern Africa*: The Juan–Fernand and Charing fault zone in the Pacific Ocean and the Rio Island-Grande fault zone in the Atlantic Ocean, etc.

(5) *Tectonic zones at 50°S–55°S of a nearly N-S-trending ocean tectonic system in the three oceans*: All tectonic zones changed to nearly E-W strikes, such as the Antarctica–Australian Ridge, Skecher Ridge and Atlantic–Indian Ocean Ridge, which is an indication of the existence of E-W-trending tectonic zones.

(6) *Tectonic zones at 60°S*: With approximately 60°S as the main axis, the Atlantic-Indian Ocean basin, the Bellingsgop sea basin, the South Pacific

ridge and the Antarctic–East Indian sea basin, etc., are all distributed in E-W-trending areas around the Antarctic continental margin.

(7) *Tectonic zones at* 70°S: There is a nearly E-W-trending fault zone along the continental margin of Antarctica. Notably, no obvious traces of the E-W-trending tectonic zones at 40°S–45°S have been found in the southern hemisphere. There are only sporadically distributed traces in the Bass Strait on the southern margin of Australia and further south and northeast of New Zealand. Ocean data show that when extending south-ward to 40°S–45°S, the medium ridge zones in the three oceans all turn sharply to both sides and form nearly E-W-trending medium ridge zones. This feature may be an indication of the E-W-trending tectonic system.[16] If so, the symmetry of distribution of the E-W-trending tectonic systems in the southern and northern hemispheres will be more apparent.

(8) *E-W-trending tectonic system in Antarctica*: As the most significant feature, this tectonic system consists of the following basins around Antarctica: the Amundsen Sea, Ross Sea, Wilkland, Puri, Xmonut Sea, Li HuiLarsen, Brontekro, Wenar Sea and South Hydran basins (Fig. 2.5).

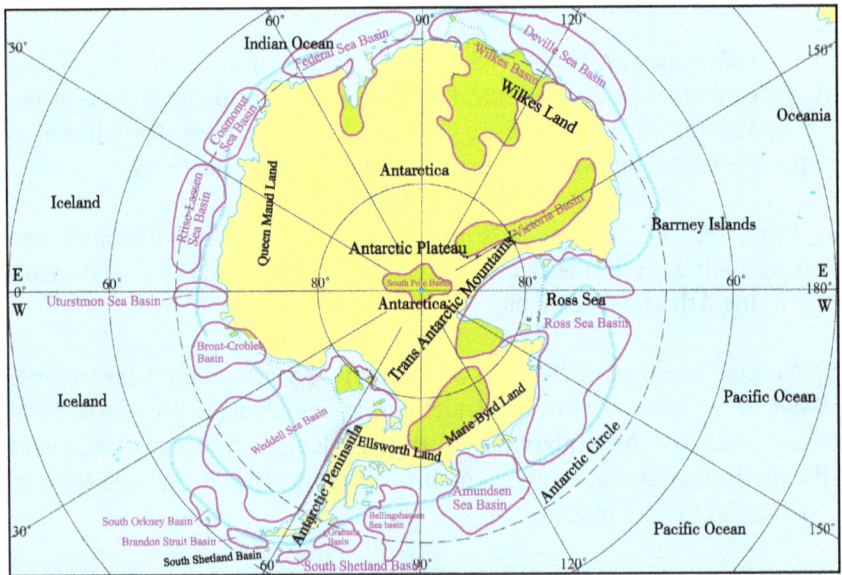

Fig. 2.5.   E-W-trending tectonic systems in Antarctica.

References

1. Li Siguang. *Introduction to Geological Mechanics*. Beijing: Science Press, 1973.
2. Li Siguang. *Vortex Structure*. Beijing: Science Press, 1974.
3. Li Shujing, Zhang Weijie. Discovery of Zonal Nappe Structure in Sunite Zuoqi, Inner Mongolia and its Geological Significance. *Journal of Geomechanics*, 1995, 1(1): 7–10.
4. Li Zikun. *Concept of Main Tectonic Systems in Anhui Province*. Institute of Geomechanics, Ministry of Geology and Mineral Resources. Collection of Studies on Provincial Tectonic Systems in China (Volume 1). Beijing: Geological Publishing House, 1985.
5. Sun Dianqing. *Structural System Map of the People's Republic of China and its Adjacent Sea Areas*. Beijing: China Cartographic Publishing House. 1984.
6. Sun Dianqing. *Geomechanics Theory and Practice in China's Petroleum Exploration*. Beijing: Geological Publishing House, 1989.
7. Wang Hongzhen. *Tectonic Paleontology and Biogeography in China's Neighboring Areas*. Beijing: China University of Geosciences Press, 1990.
8. Tian Zaiyi. Analysis of Oil and Gas Prospect in Junggar Basin from Geological Development History. *Xinjiang Petroleum Geology*, 1989, (3): 3–14.
9. Bao Ci, Yang Xianjie, Li Dengxiang. Geological Structural Characteristics and Natural Gas Prospect Prediction in the Sichuan Basin. *Natural Gas Industry*, 1985, 5(4): 1–11.
10. Yuzhu Kang, Zongxiu Wang *et al*. *Study on Oil Control Effects of Sichuan Basin Tectonic System*. Beijing: Geological Publishing House, 2014.
11. Yuzhu Kang, Zongxiu Wang *et al*. *Study on Oil Control Effects of Geological Structure System in Northwest China*. Beijing: Geological Publishing House, 2013.
12. Apak, S. N., Stuart, W. J., Lemon, N. M. Structural Stratigraphic Development of the Gidgealpa-Merrimelia-Innamincka Trend with Implications for Petroleum Trap Styles, Cooper Basin, Australia. *Journal of the Australian Petroleum Exploration Association*, 1993, 33(1): 94–104.
13. Asami, M., Suzuki, K., Grew, E. S. Monazite and Zircon Dating by the Chemical Th-U-Total Pb is Ochron Method (CHIME) from Alasheyev Bight to the Sor Rondane Mountains, East Antarctica: A Reconnaissance Study of the Mozambique Suture in Eastern Queen Maud Land. *Journal of Geology*, 2005, 113: 59–82.
14. Askin, R. A., Elliot, D. H. Geologic Implications of Recycled Permian and Triassic Palynomorphs in Tertiary Rock of Seymour Island, Antarctic Peninsula. *Journal of Geology*, 1982, 10: 547–551.

15. Yuzhu Kang. *Relationship Between the Tectonic System and Oil and Gas in the Tarim Basin*. Institute of Geomechanics, Chinese Academy of Geological Sciences. Collection of Geological Mechanics (Part IX). Beijing: Geological Publishing House, 1989.
16. Yuzhu Kang. Discussion on Oil and Gas Distribution Law and Oil Exploration Direction in Tarim Basin. *Earth Science*, 1991, 6(1): 429–436.

Chapter 3

N-S-Trending Tectonic Systems

**Yuzhu Kang, Shuwen Xing, Yue Zhao, Yinsheng Ma,
Dewu Qiao, Zongxiu Wang, Xingui Zhou,
Zhihong Kang, and Zhihu Ling**

Abstract

In this chapter, N-S-trending tectonic systems are introduced, including those in the Ural Mountains in Asia, the island of Sakhalin in Russia, the Andes Mountains in South America, East Africa, the oceans and China.

Keywords: Tectonic system; N-S-trending; Type

The main body of the N-S-trending tectonic system stretches in the N-S-trending direction. The system is composed of a group of N-S-trending compression tectonic zones or tension-fault tectonic zones, accompanied by the N-W-trending zone, N-E-trending torsion zone and nearly E-W-trending expansion zones. Each tectonic zone group forms its own system, which is the product of the E-W-trending zone squeezing and stretching the Earth's crust. At the same time, the N-S-trending system features large-scale and corresponding deformation zones, accompanied by corresponding magmatic activity, deposition formations, and tectonic dynamic metamorphic zones. They affect the Earth's crust at different depths. A large impact depth reflects more obvious gravity and magnetic cascades and basic and ultrabasic magmatism. In contrast, a small impact depth reflects insignificant magmatism caused by slips in the crust surface or shallow layers and regional features, e.g., some N-S-trending tectonic zones south of the Chinese continental crust. However, regardless of the type or level of N-S-trending tectonic zone, the distribution in the upper crust is approximately equidistant, and the mechanical properties tend to be regional. Especially for the Ural N-S-trending compressional tectonic systems in central Eurasia and the N-S-trending mountain systems between the western margin of the American continent and the eastern coast of the Pacific Ocean, there are many differences in scale, strength and impact depths, but the systems are basically compression oriented. In contrast, the N-S-trending tectonic systems in Europe, Africa and the Atlantic Ocean and on both sides of the Urals are tension oriented, i.e., the East Africa rift zone, Dead Sea in western Asia, Jordan Valley, Rhone Valley, Rhine Valley, Scandinavian Fault and Taixi Rift Zone in western Europe. This characteristic may have occurred during the crustal movement caused by the Earth's rotation, when the Asian continent advanced more quickly than the American continent. The Asian continent and the eastern coast of the Pacific Ocean were squeezed in the E-W direction, while the European and African continents and the Atlantic Ocean were torn apart.[1]

The N-S-trending tectonic system is one of the most basic and common tectonic systems in the Earth's crust. It exists everywhere in the continents and oceans, and some N-S-trending tectonic systems could form large N-S-trending tectonic systems.

3.1 Ural N-S-Trending Tectonic System in Asia

The main body of this system is located at 50°E–62°E, starting from Xindi on the north and extending to the southern shore of the Aral Sea on the south, with a length of more than 2,000 km. Moreover, it forms the boundary between Europe and Asia and a Cenozoic trough distributed at approximately 60°E south of the Aral Sea.

The N-S-trending tectonic trace is still very significant in the contiguous zone of Afghanistan and Iran, so it was once called the Ural–Oman tectonic zone and is now generally called the Ural fold system. The western section is an orogenic zone, and the main body comprises the Ural Mountains and the Mugozare Mountains, with a total length of approximately 2,500 km. The eastern and southern sections are hidden under the deposition caprock in the Ob Rivers Basic–Tulan Depression Region. The entire fold system measures approximately 4,000 km long with a local width of 500 km. It is composed of a series of N-S-trending linear fold zones and fault zones. The Late Precambrian–Cambrian strata are the oldest strata, forming the Ural tectonic zone, and all tectonic elements earlier than the Late Precambrian have different directions from the Ural N-S-trending tectonic zone. Therefore, the formation controlled by the Ural tectonic zone was developed at the end of the Proterozoic, and its latest deposition strata constitute mainly the Triassic system, with a local Lower Jurassic system. The tectonic zone can be divided into (1) West Ural marginal tectonic zone (asymmetrical fold zone formed by Ordovician–Triassic strata); (2) Middle Ural compound anticline zone, which is covered by the Cenozoic system on the west and gabbro peridot intrusions and Ordovician-early Devonian eruptive rocks on the east; (3) Tagil–Magnitogor compound syncline zone, which is composed of the Proterozoic-Carboniferous and widely distributed intrusive rocks; (4) East Ural compound anticline zone, which is composed of mainly the Middle Paleoproterozoic and extensively developed Hercynian acid intrusive rocks; (5) Alapayev Bred syncline zone, a linear fold zone formed at the end of the Carboniferous and a Graben tectonic zone formed along the Ural N-S-trending zone and filled with Permian and Triassic sediments; (6) Tobol compound anticline zone, which formed mainly at the end of Carboniferous in the nearly meridian direction, and its eastern section is a Silurian and Devonian depression zone; (7) Valeryyanov syncline and Borov anticline zones, which are completely hidden under Cenozoic conditions and determined on the basis of geophysical and drilling data and are composed mainly of the Carboniferous N-S-trending Ural fold zone and partial grabens filled with Permian and Triassic sediments. In addition, the eastern section was sometimes filled with active pre-Carboniferous sediments in the Kazakhstan fold zone.

The faults in the Ural fold zone are well developed, with deep faults in the upper mantle controlling the tectonic lithofacies subzone in the Ural trough. They are also related to the formation of ultrabasic rock zones.[2] Such faults were distributed mainly between depressions and relatively uplifted zones. In Ordovician–Early Triassic strata in the western Ural marginal tectonic zone, a series of overturn faults and reverse faults developed. In contrast, in the western and eastern parts of the Ural Mountains, some large and flat overturn faults developed, which were mainly Cenozoic tectonics.

The Ural N-S-trending tectonic system has experienced long-term and multiple tectonic movements. From the end of the Neoproterozoic to the Early Paleozoic, it was basically a N-S-trending trough deposited with turbidite (flysch) deposition rocks, carbonate deposition rocks and submarine volcanic eruptive rocks that together were tens of thousands of meters thick. In the Late Paleozoic–Early Mesozoic, the folds were uplifted and accompanied by a large amount of intermediate-acid magma intrusions. The central and southern sections of the zone were merged by the spine of the Eurasian epsilon-shaped tectonic.

3.2 Sakhalin N-S-trending Tectonic System in Russia

The system spreads at approximately 140°E–145°E, starting from north of Sakhalin in the north and passing through the Hidaka Mountains of Hokkaido, North Uphill, Izu Islands, the Tartar Strait of the island of Honshu, the Japanese Trench, the Mariana Islands and the famous Mariana Deep Trench in the south. After being crossed by the E-W-trending Caroline Islands, there are still traces at approximately 142°E, up to near the equator. After crossing the middle of Irian Island, the system is intermittently connected with the Cape York Peninsula in Oceania, forming a giant N-S-trending tectonic zone on the margin of East Asia or the western margin of the western Pacific. The zone exhibits the same features as Koryak and the Kamchatka Thousand Islands in modern tectonic movements and magmatic activities, as well as manifestations since the Late Jurassic. Because this zone is located on the western side of the West Pacific tectonic zone and is close to the East Asian continent, the activities in the Sakhalin–Hokkaido region are weakened, and the fold movements and magmatic activities are also relatively weakened. There are three Paleozoic and Mesozoic compound anticlines in the eastern, southern and Rishan sections of Sakhalin, two Paleozoic and Lower Triassic compound anticlines in the Beishangshan and Awudian regions, and Mesozoic and Cenozoic N-S-trending tectonic magmatic zones composed of nearby N-S-trending Neoid volcanic rock zones, Late Mesozoic acid intrusive rocks, and Late Mesozoic–Paleogene basic and ultrabasic rocks.

The strong fold deformation of the Sakhalin N-S-trending tectonic zone occurred mainly at the end of the Cretaceous and then experienced deformation and metamorphism, such as folds and faults, at the end of the

Pliocene, integrating the zone and the Xihuote Mountains into a unified continent. At the end of the Quaternary, it sunk along the N-S-trending fault, forming the Tartar Strait and Sakhalin.

Japan is an important part of the Mesozoic and Cenozoic trough systems in the Pacific. Because Japan is located at the boundary between the continental crust and the oceanic crust, it is one of the most active zones on Earth today, with island arcs and marginal trenches and the greatest depth and the largest gravity anomaly in the oceans, featuring extensive volcanic activities and strong seismic activities. From a tectonic point of view, there is a N-S-trending fault zone in the middle of the Japanese island arc called the Itoigawa–Shizuoka fault zone. The southern Izu–Ogasawara intra-arc zone and Mariana intra-arc zone are the Neogene "green tuff" distribution zones. In contrast, the adjacent Izu–Ogasawara extra-arc zone and Mariana extra-arc zone are not the Neogene "Green Tuff" distribution zones but porphyry basalt distribution zones formed at approximately 140°E. The Izu–Ogasawara zone is the most active deep earthquake zone, responsible for 67% of deep earthquakes in Japan.

3.3 Andes N-S-Trending Tectonic System in South America

This N-S-trending tectonic system is the largest N-S-trending compression tectonic zone in the Americas, starting from southwest of the Caribbean Sea on the north, extending southward along the Andes Mountains, and forming an arc-shaped tectonic protruding westward or eastward at approximately 75°W. Its northern section is dominated by the eastern and western sections of the Cordillera Mountains, forming an arc-shaped tectonic belt protruding westward, with an arc-shaped top at approximately 80°W. The Andes Mountains are S-shaped. The system terminates at the Antarctic Ridge in the south and spans approximately 70° from north to south. It is composed of synclines and anticlines in the Paleozoic. It partially transformed into a strong fold-fault zone at the end of the Paleozoic and an Andean-Cordillera N-S-trending fold zone at the end of the Mesozoic and was still active during tectonic movements.

Notably, the Cordillera Mountains on the western coast of the Americas are the junction zone between the North American continent and the eastern margin of the East Pacific. They have been squeezed and twisted between the Pacific and American blocks since the Late Mesozoic, especially in the Neoid period. However, there is a compression-torsional tectonic zone characterized by clockwise strike-slip in the region from north of the Caribbean Sea to Alaska. Furthermore, it formed a large-scale reverse S-shaped tectonic system, which contained or merged some fragments of nearly N-S-trending tectonic zones. The relationship between the N-S-trending tectonic zones and the Cordillera–Andes N-S-trending tectonic zone on the western margin of South America needs to be further studied.[3] Therefore, the Andean N-S-trending tectonic system referenced here comprises mainly the Cordillera Mountains south of the Caribbean Sea.

3.4 N-S-Trending Tectonic System in East Africa

This system is mainly distributed at 30°E–40°E. It is composed of a series of extensional rift zone groups and is a N-S-trending extensional tectonic zone commonly known as the East African Rift Zone. From the perspective of its distribution, it can be approximately divided into two branches (the eastern branch and the western branch). The western branch starts from Abbott Lake in the north and passes through Edron Lake, Kivu Lake, Tanganyika Lake and Rukwa Lake. The eastern branch is called the Grigori Rift Zone; it starts from Ludolf (Turkana Lake) in the north and passes through a series of small lakes and Natron Lake. The two branches converge at Niassa Lake and extend to the lower reaches of the Zambezi River. A single rift zone has a width of approximately 50 km and is composed of a series of nearly N-S-trending steeply dipping normal faults, with a vertical fault distance of 1,000–2,000 m. The southern section of the system formed in the Carboniferous and was active in the Jurassic and Cretaceous. The entire tension zone formed in the Cenozoic. In addition to tensional activities, there are torsion signs. This large rift zone measuring thousands of kilometers long is controlled by two sets of torsion planes at its margins and spreads and turns in the overall N-S-trending direction, forming this N-S-trending extensional fault tectonic zone. It is the most typical and largest N-S-trending extensional fault zone on the continent.

The zone corresponds to the Dead Sea–Jordan Valley N-S-trending tectonic zone to the north at 34°30'E–37°E. Based on available data, this section is a large-scale rift zone formed from the beginning of the Cenozoic to Oligocene on the basis of the N-S-trending compound anticline and thrust zones formed by the strong compression of the Paleozoic and Mesozoic strata, in which the Pleistocene continental plateau basalt zone is approximately 1,000 km long. In addition to its tensional features, it is also twisted counterclockwise. Therefore, the Dead Sea–Jordan Valley N-S-trending tectonic zone may be the northern extension of the N-S-trending tectonic system in East Africa, and both areas are connected in the Red Sea Valley. The Red Sea Valley extends from the Mande Strait to the southern end of the Sinai Peninsula in the N-N-W direction, with a length of 1,930 km. The Red Sea Basin appeared before the Oligocene. When it began to sink, it was a valley with a chain of freshwater lakes. In the Miocene, seawater intruded into the depression, and gypsum-bearing limestone was deposited. These data indicate that the N-S-trending fault tectonic system in East Africa formed mainly since the Oligocene, developed rapidly in the Miocene and Pleistocene and is still active today.

3.5 N-S-Trending Tectonic System in the Oceans

Marine survey data show that mid-ocean ridges are common on the floor of the three major oceans on Earth, and they are nearly N-S-trending. Due to the misalignment of the E-W-trending tectonic zones, the mid-ocean ridges are often curved in an S-shape. The seabed landforms of mid-ocean ridges are wide and gentle underwater uplift zones, so they are also called ocean ridges. There are often narrow and long valleys (rift valleys) or small ridges and troughs alternately in the middle of the uplift zones. Moreover, there are high heat currents along the axis of the ocean ridges, and they are weakened significantly on both sides. Volcanic rocks at the axis of the uplift zones are the youngest in age, and they are older on both sides. In general, the mid-ocean ridge is tensioned and sometimes twisted.

The main N-S-trending ocean zones include the Pacific Mid-Ridge, Atlantic Mid-Ridge, Central Indian Ocean Mid-Ridge and East Indian Ocean Mid-Ridge (90° ridges). Among them, the first two are larger, while the latter two are smaller in scale. The zones extend southward to approximately 50°S and are all E-W-trending, forming E-W-trending rift zones surrounding the Antarctic continent. Due to the misalignment of a series of E-W-trending tectonic zones and E-W-trending transition faults, the various sections of each ridge are not located at the same longitude, and some nodes are dislocated and intermittently formed into a zone, but the overall connection is N-S-trending. The Atlantic Mid-Ridge extends northward into the Arctic Circle and is called the Arctic Mid-Ridge. Its distribution is parallel to and equidistant from the Lomonosov Strait and Mendeleev Ridge. The Central Indian Ocean ridge extends at 70°E–75°E, and its middle section is in the N-S-trending direction with a large width. The most typical section is the Maldives–Chagos Islands, also called the Maldives Ridge. The East Indian Ocean ridge spreads at 90°E, so it is also called the 90° Ridge. It is a straight N-S-trending underwater uplift zone, which became twisted clockwise in the Late Cretaceous.[4]

The N-S-trending oceanic rift tectonic system features are as follows: (1) they have obvious orientations, and all rift zones are generally N-S-trending, which is basically parallel to the meridian direction; (2) they are large in scale, i.e., thousands to tens of thousands of kilometers in length, tens, hundreds or even thousands of kilometers in width, and thousands of meters high above the seabed; (3) the rift zones are composed of longitudinally extended tensional faults, with the steep walls being inclined to the middle and the rifts in the middle; (4) perpendicular to the rifts, there are lateral conversion faults, which are actually sets of matched compression and torsion faults; (5) a large number of volcanic rocks are in the rifts, so-called mantle plumes are formed in some places, and the heat flow value is generally high; (6) under tension effects, blocks or rock blocks on both sides of the rifts expand horizontally, forming giant expansion zones; (7) the four rift zones are evenly divided into two branches in the Southern Hemisphere, whose rifts together form a diamond-shaped rift zone surrounding the Antarctic continent; (8) the Tasman rift zone has been transformed from a tension to a compression state; (9) the strata on both sides of the rift zones are asymmetric, i.e., newer near the rift zones and older on both sides, and the oldest strata were formed in the Jurassic; and (10) the distribution of the four rift zones on the Earth is equidistant, with a spacing of approximately 90° longitude.

The N-S-trending tectonic systems have the following features. All tectonic zones are developed unevenly and asymmetrically in the Northern and Southern Hemispheres. In terms of their mechanical properties, the N-S-trending tectonic systems are dominated by compression, are well developed and are widely distributed in the Eastern Hemisphere. In contrast, they are dominated by a group of tensional and tension-torsional tectonic zones with a wide distribution in the Western Hemisphere and mostly tensile tectonic zones developed by tracking two sets of torsional faults, i.e., the East African Rift Zone, Rhine Graben Zone and Atlantic Tension Rift Zone. Overall, they are subject to strong E-W-trending compression (sometimes with torsion) in the Eastern Hemisphere and the E-W-trending tension and tearing effect in the Western Hemisphere, and these activities have been more regular since the Mesozoic. From a global perspective, the main N-S-trending giant squeezing and extensional tectonic zones feature direction, positioning and equidistance. From the perspective of the N-S-trending squeezing tectonic systems on the Eurasian block, they feature equal intervals of 40°, 20° and 4° longitudes. They are curved in the direction or misaligned by the nodes, but they spread basically in the nearly N-S-trending direction. They are composed mainly of N-S-trending folds, thrusts and squeezing fault zones, tectonic dynamic metamorphic rock zones, magmatic rock zones and corresponding deposition formations, accompanied by N-E-trending, N-W-trending and nearly E-W-trending extensional fault zones, as well as derived series of secondary torsion tectonics. They are generally large in scale, with a length of more than 1,000 km, a width of tens to hundreds of kilometers and a large depth. Sometimes they spread to the upper mantle, and gravity, aeromagnetic and other anomalies appeared in these zones. For example, the Ural zone, Sanjiang zone, Sichuan–Yunnan zone, Chinese Taiwan–Philippines zone, Sakhalin–South Islands zone and other N-S-trending tectonic systems have obvious control effects on the formation and distribution of deposition formation, magmatic activities, dynamic metamorphic zones, and geophysical and geochemical anomaly zones. Some zones have a long development history, with good development in the Paleozoic, Mesozoic and Cenozoic. There are N-S-trending compressional tectonic zones in many pre-Paleozoic blocks. Some fault zones merged and transformed into the components of the later N-S-trending tectonic zones, while the ancient N-S-trending fold zones were often reconnected and contained in the later N-S-trending zones, but it is difficult to determine the orientations of their formation. This phenomenon is clearly shown in the East Hebei–Zhangbaling zone, the Mudanjiang–Tumen River zone, the Sichuan–Yunnan zone and Sanjiang zone, as well as in the Ural zone in Russia, the Baltic Sea block and the Yiergang block. The giant N-S-trending extensional tectonic zones in the continents and oceans have both similarities and differences. They all take the form of fault depressions, basins, troughs, etc., with high heat flow values and significant geophysical anomalies, but the rift zones differ in terms of controlling deposition formation, magmatic activities and mineralization from the oceanic rift zones.

There are many hypotheses about the formation mechanism of the significant differences in the mechanical properties of the N-S-trending tectonic zones in the Eastern and Western Hemispheres. Based on research on tectonic dynamics and kinematics caused by the rotation of the Earth in geomechanics, the main reason for this formation is believed to be that the Eurasian continent was torn apart at the different slip speeds of the continental blocks (the Asian continent advanced forward, while the American continent retreated backward) under the action of the Earth's rotation, as stated by Li Siguang: "Asia has advanced, and America has retreated, while Europe has been torn apart, and Africa has been finalized". The positioning and orientation of the N-S-trending tectonic systems may be caused by the N-S-trending coordination function zone created by changes in the Earth's rotation speed.[5]

3.6 N-S-Trending Tectonic Systems in China

The N-S-trending tectonic zones on the Chinese continental crust are composed of N-S-trending strongly compressional tectonic zone groups and dominated by single or multiple severe folds or fold groups. In some regions, there are often large N-S-trending compression or torsional fault zones, sometimes accompanied by strong tectonic dynamic metamorphic zones and tectonic magma zones, forming complex tectonic magma dynamic deformation and metamorphic zones, such as the Sanjiang (which comprises Nujiang River, Lancang River, Jinsha River) N-S-trending tectonic zone and the Sichuan–Yunnan N-S-trending tectonic zone in western China and the Mudanjiang–Tumenjiang N-S-trending tectonic zone in eastern China.

The N-S-trending tectonic systems are most prominent in southern and southwestern China. In northern and western China, except for a few tectonic zones, they are generally scattered and not very strong. They are often combined with some components of other tectonic systems. In general, the N-S-trending tectonic zones are widely distributed throughout the continental crust of China, and some zones have good continuity and pass through the Chinese mainland, i.e., the Sanjiang zone, the Sichuan–Yunnan zone, the East Anhui zone and the Mudanjiang zone.

The extensions of most N-S-trending tectonic zones exhibit low intensity and do not correspond in the southern and northern directions. For example, they are distributed widely on the border among Shaanxi, Gansu and Ningxia, both sides of the Shanxi continental platform, Daxinganling, Guizhou Plateau, Hunan, Guangdong, Jiangxi and Taiwan. In recent years, they were discovered in the northwestern region and Tibet, but most of them are weak compression tectonic zones. Furthermore, due to the separation and right misalignment of the Qinling–Kunlun E-W-trending zone and the Tianshan–Yinshan E-W-trending zone, they appear in sections in space, and the southern, middle and northern sections are misaligned, so it is difficult to correspond in space. Their N-S-trending correspondence seems to have changed mostly in different geohistorical periods. For example, based on geohistorical data, the Sichuan–Yunnan N-S-trending tectonic zone may have belonged to the same tectonic zone as the Helan Mountain zone in its early periods, i.e., Jinning and Caledonian periods, and is an important tectonic magma activity zone that affects the developmental history of the eastern and western parts of the Chinese continental crust. However, from the Late Paleozoic to Yanshanian period, they belonged to two N-S-trending tectonic zones. The Sichuan–Yunnan zone extended to the north and connected to the Qinghai–Tongren zone via the Xiqing Mountains, while the Helanshan zone was consistent with the Sichuan–Guizhou N-S-trending zone in terms of genetic development and spatial distribution.

Notably, if the rock strata (N-S-trending) that form the strong and repeated fold-fault zones in southwestern and southern China are flattened, they cannot be accommodated within the region where they are located. Therefore, these strata are limited to the surface or the upper part of the crust. They were originally flattened in the farther east region, and after horizontal slips to the west, they formed the current fold zones. The earthquakes in the southwestern region were often shallow earthquakes, which reflected the horizontal slip of the upper crust (Li Siguang, 1973). Since the Mesozoic and Cenozoic, some N-S-trending tectonic zones have intersected the E-W-trending zones and other early tectonic zones, showing good continuity. Therefore, some sections of the same N-S-trending system contained and merged some early rock zones and rock blocks, but it is difficult to determine early positions and orientations today.[6] Whether there are corresponding tectonic zones in different blocks can be determined by fully considering the tectonic evolution and development of different blocks.

3.6.1 Sanjiang N-S-Trending Tectonic System

This system spreads at 90°E–100°30'E and is distributed roughly along the Jinsha River, Lancang River and the middle reach of the Nujiang River in western Yunnan and eastern Tibet. Hence, it was originally named the western Yunnan N-S-trending tectonic zone. It is the main site of the famous Hengduan Mountains in Southwest China. Along the zone, there are steep ridges, peaks and mountains, as well as turbulent rivers flowing from north to south. The system extends southward into Myanmar and western Thailand. Due to the effects of the Qinghai–Tibet–Sichuan–Yunnan reverse S-shaped tectonic zone, its longitude shifted approximately 2° westward in western Thailand. The main body appeared west of the Chao Phraya River and east of the Andaman-Nicobar Islands, forming an arc-shaped zone protruding westward. This arc-shaped zone featured strong activity in the Cenozoic. The main body of the system extends to the Queer Mountains north and east of the Bayan Har Pass and is connected intermittently to the N-S-trending tectonic zone in the Ulan and Dulan regions of Qinghai.

The Sanjiang N-S-trending tectonic zone is divided into three parts or three subzones by the Nujiang fault zone and the Lancangjiang–Kejie fault zone. Although all parts have distinct geological tectonic features, they may not have the same crystalline basement or the same tectonic development history. Due to the influence and interference of the Kunlun E-W-trending zone, the N-W-trending system, and the Qinghai–Tibet Sichuan–Yunnan inverse S-shaped tectonic zone, it is discontinuous and is divided into three sections (the southern section, the middle section and the northern section). The middle and southern sections are divided by the Shuanghu–Dingqing–Ailaoshan fault zone, which forms an important boundary line between the Yangtze–Talimu block and the Cathaysia–Indosinian block. The Shuanghu–Dingqing section is the boundary line between the Yangtze–Talimu block and the northern margin of the southern Tibetan block. Significant commonalities appeared on the northern and southern sides after the Caledonian movement, and this N-S-trending tectonic zone formed in the Late Hercynian and Early Indosinian. In the Yanshan-Himalayan period, due to the composite superposition of the Qinghai–Tibet–Sichuan–Yunnan reverse S-shaped tectonic system and the activities of the zone itself in the Neoid period, it was subject to strong compression and clockwise strike-slip, forming a geological and geomorphic landscape through southwestern China today. The southern, middle and northern sections are briefly introduced as follows (Fig. 3.1).

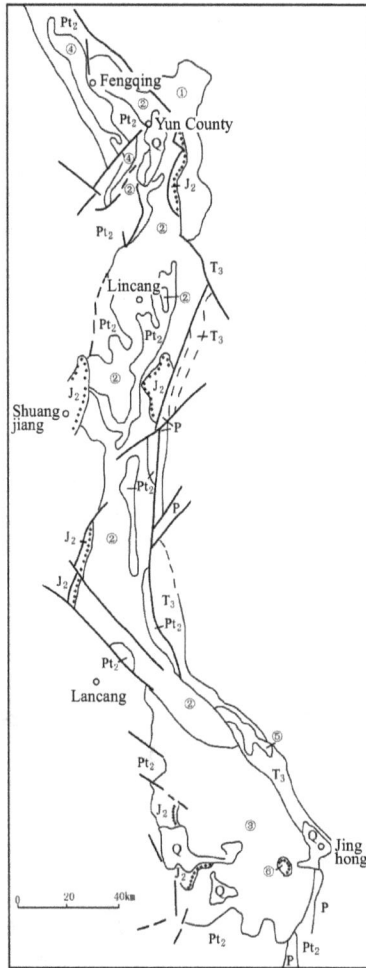

Fig. 3.1. Geological diagram of Lincang compound rock foundation.

The main body of the southern section spreads in the western Yunnan region between the Lancangjiang fault zone and the Nujiang fault zone, extending northward into western Sichuan and eastern Tibet and southward into Myanmar and western Thailand and then the region between the Andaman and Nicobar Islands in India.[7] It is divided by the Nujiang fault zone and the Kejie fault zone into eastern, middle and western subzones, which are called the East Asian Zone–Changning–Menglian fold-fault zone, Central Asia Zone–Fubei–Zhenkang fold-fault zone and West Asia Zone–Bosulaling–Gaoligong Mountain fold-fault zone, respectively. The easternmost boundary is the South Lancangjiang fault zone, which is an important tectonic magma dynamic metamorphic zone that separates the Cathaysia–Indosinian block from the Nanqiangtang block and is called the Taima–Lancangjiang suture zone. Although these three zones have their own geological tectonic features, they have a similar crystalline basement and approximately the same tectonic evolution process, so they should be considered components of the same tectonic system.

There are nearly 100 known rock masses, including ferrous rocks, mafic-ultramafic rocks and diabase masses. The gabbro and ultramafic rocks are large in scale and closely follow each other, intrude into the Permian and Upper Triassic systems, and are distributed in the direction of the regional tectonic line. The mafic-ultramafic rocks intruding into the Lower Permian system are vertical to the subzone. The diabase masses are widely distributed and have intruded into the Permian-Jurassic system in multiple stages.

The development of a high-pressure dynamic metamorphic rock zone is another important feature of the East Asian zone. Especially for the Damang Guangfang dynamic metamorphic zone, there are high-pressure dynamic metamorphic minerals, i.e., blue amphibole-bearing plagioclase mud schists, amphibole-bearing sericite blue schists, polysilica-bearing muscovite sodium-sodium sericite schists, etc., in the Lancang Group green schists along the fault zone and the Lower Carboniferous basic volcanic rocks near Damang Guangfang in Shuangjiang County. The blue amphibole-bearing rocks in the Lancang Group are generally broken and schistosized, while the blue amphiboles are complete, and some are produced in the mineral fissures and obliquely intersect with the inner walls at a certain angle. These features indicate that they were formed in the breakage stage of the existing rocks or later, and they also appeared in the Lower Carboniferous basalts. The measurement of polysilicon muscovites with the Rb-Sr method shows that their age value is 240–260 Ma. For the Changning–Menglian metamorphic rock zone as a whole, the latest layer is the Upper Permian system, above which there is an unmetamorphic Triassic system. The measurement of Gengma granites intruded in metamorphic rocks with the zircon U–Pb method shows that its age is 223–241 Ma. These measurements indicate that the main metamorphism occurred in the Late Hercynian–Early Indosinian, which is consistent with the deformation period and the main intrusion period of regional magmatic rocks. The deformation and metamorphism in this period are closely related to the superposition and transformation of thrusting and overthrusting in the same period or later periods.

The corresponding main faults in the southern section of the Sanjiang N-S-trending tectonic zone include the southern section of the Lancangjiang fault zone, the Jinggu–Jinghong fault zone, the Damang Guangfang–Lancangjiang fault zone, the Kejie fault zone and the southern section of the Nujiang fault zone.[8] Except for the eastern and western branch faults of the southern section of the Nujiang fault zone appearing on the West Asian zone, other major fault zones appear on the eastern and western margins and in the center of the East Asian zone. The N-S-trending tectonic zone extends through Ulan, Dulan and West Yangkang and enters the North Qilian zone at Ulan Daban.

The above descriptions show that the Sanjiang N-S-trending tectonic system has good continuity and connectivity. It extends northward and passes west of the Badain Jaran, corresponding to the Daalhat fault, Kusubo fault and other large basement faults in Mongolia. Its southern section is connected with the Thailand–Chao Phraya–Middle Malaysia junction zone and the N-S-trending tectonic zone on the western side. The main deformation and metamorphism periods of this system are the Late Hercynian and Early Indosinian, while the N-S-trending deformation and metamorphism are not obvious in the Early Paleozoic. However, the nearly N-S-trending depression zone seems to have appeared, forming a component of the Qingkang–Dian–Burma trough, which may have had a certain control effect on the magmatic activities in the Late Caledonian. A mafic-ultramafic rock zone appeared along the large fault zone from the Late Paleozoic to Indosinian. From the Late Indosinian to Early Himalayan, this zone was affected and superimposed mainly by the Tethyan tectonic zone and was compounded by the Qinghai–Tibet–Sichuan–Yunnan reverse S-shaped tectonic system. There was a significant clockwise strike-slip in the middle-southern section of its main fault zone in the Late Himalayan. Moreover, due to the Qinghai–Tibet Plateau uplift and E-W-trending compression, there was an E-W-trending thrust nappe and a clockwise strike-slip in the Sanjiang N-S-trending tectonic zone, resulting in obvious tectonic displacement and deformation in the region and more complex tectonic changes.

The Sanjiang N-S-trending tectonic zone is a Level I gravity and magnetic feature line group. In the Bayan Har Mountain region, the magmatic rock zone along the Rubber Mountain–Animaqing fault zone and

Jiangda–Maduo fault zone is a Level II gravity and magnetic feature line group.

These two feature line groups also extend northward continuously like the surface, pass through the East Kunlun tectonic zone and enter the western side of Qinghai Lake. On both sides of the uplift zone on the western side of Qinghai Lake, a set of heavy and magnetic overlapping feature lines extends from Maduo and the main peak of the Qilian Mountains into the Gansu mountainous region, with a length of more than 500 km. The overall combination of Bouguer gravity, regional magnetic field and linear anomaly zones in western Yunnan converge to the north and spread to the south. The distribution and combined features of the gravity and magnetic anomaly zones are consistent with those of the main fault zones and tectonic magmatic metamorphic zones east of the Nujiang River and in this region.

3.6.2 *Sichuan–Yunnan N-S-trending Tectonic System*

This system is also known as the Sichuan–Yunnan N-S-trending tectonic zone. It is composed mainly of two basic tectonic units of the Kangdian anticline or the Kangdian earth axis and the Liangshan–Diandong depression zone on the eastern side and is distributed mainly at 101°30'E–140°E. The N-S-trending fold mountain system in Mount Gongga, Daxue Mountains, Daxiaoliang Mountain and Min Mountains of western Sichuan and central Yunnan has the most developed tectonic features. Its northern section is suppressed by the Ruoergai–Red Plain Arc zone, but there are strong traces in Minshan and Ruoergai, as well as in East Qinghai Lake–West Lanzhou and the Tengger–Badain Jaran Desert contiguous region after the Qinling Mountains. The southern section has strong manifestations south of the Ailao Mountains, western Vietnam, and Laos and on the eastern coast of the Malay Peninsula. The West Sichuan–Central Yunnan section of this system has been studied in detail with abundant data and can be divided into three subzones from west to east: the western tectonic magmatic metamorphic zone, the central uplift zone and the eastern fold-fault zone (Figs. 3.2 and 3.3).

3.6.3 *Hunan–Guangxi N-S-trending Tectonic System*

This system is distributed near 109°W–113°E, is well developed in southern Hunan and northeastern Guangxi and extends to northwest

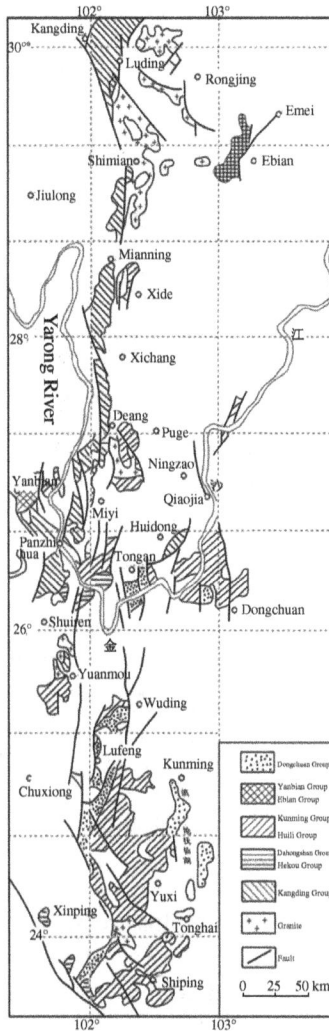

Fig. 3.2. Distribution of the pre-Sinian system in the Xichang–central Yunnan region of China.

Guangdong to the south. It is more scattered than the above N-S-trending tectonic system and has a length of 200 km. This system is composed of mainly a series of N-S-trending fold groups and compressional fault zones, as well as some small granite masses and small Jurassic and Cretaceous basins, and intersects obliquely with the N-E-trending torsion faults and the N-W-trending torsion faults and vertically with the nearly

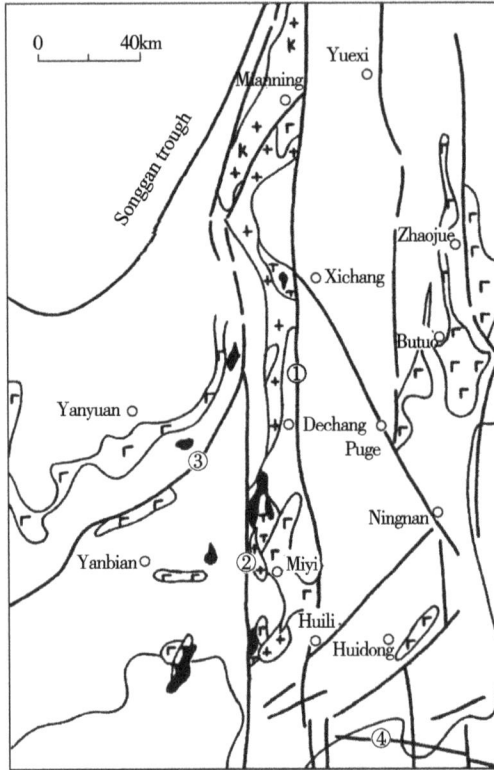

Fig. 3.3.   Distribution of post-Cambrian magmatic rocks in the Xichang region of China.

E-W-trending tensional faults. The folds are mostly broad and gentle anti-clines and narrow synclines and feature short-axis box-shaped trough-like folds that do not extend far and follow each other intermittently. The Hunan–Guangxi N-S-trending tectonic system can be divided into three secondary tectonic zones from west to east as follows: the Qianyang–Xiangchuan fold-fault zone, the Daoxian–Zhaoping fold-fault zone and the Leiyang–Linwu fold-fault zone.[9]

(1) Qianyang–Xiangzhou Fold-Fault Zone
The zone is exposed at approximately 110°E and is composed of large high-angle thrust faults, some folds and granite rock masses, with a length of 340 km, a width of 20 km and a southern tip as wide as 40 km. The main fault zone is the Longsheng–Yongfu nearly N-S-trending deep fault

zone, extending northward to the Qianyang region of Hunan. It is a multistage active fault zone developed on the basis of the nearly N-S-trending paleo-fault zone formed since the Xuefeng (Jinning) Movement and has obvious control effects on the Neoproterozoic, Sinian and Paleozoic deposits and magmatic activities. A series of positive and negative aeromagnetic anomalies are distributed along the fault zone and show a gravity gradient zone with a "high-on-west and low-on-east" feature. The main fault is a west-dipping reverse fault accompanied by east-dipping positive faults. There are several thousand-meter-wide fault-breaking mylonitization and schistization zones in the Tongwan region of Wuxuan County. The compound folds are also well developed in this fold-fault zone. The northern section is composed mainly of the Xinning synclines formed in the Upper Paleozoic and Lower Triassic, on which the Xinning Cretaceous basin was superimposed. In the southern section, the Dayaoshan secondary fold-fault zone is the most obvious and extends to the eastern wing of the Guangxi Epsilon-shaped front arc zone, represented by the Dayaoshan (Jinxiu) N-S-trending anticlines. In the Liuzhou–Xiangzhou–Laibin region, the Upper Paleozoic nearly N-S-trending widespread folds and faults are also well developed.

(2) Daoxian–Zhaoping Fold-Fault Zone

The zone spreads at 110°50'E–111°55'E and is composed of nearly N-S-trending compound folds and compressional faults, with lengths of more than 200 km and widths of approximately 100 km. It can be approximately divided into the eastern and western subzones. The western subzone is the Haiyangshan–Huashan fault uplift zone located at 110°50'E–111°25'E and is composed of the Yangshan–Gongcheng–Yuankou–Fuchuan nearly N-S-trending compound folds, Limu–Majiang and Fuchuan–Xiwan faults, some granite masses characterized by short-axis and dome-shaped folds, and the developed same-trending thrust faults, as well as some small Jurassic and Cretaceous basins. The Limu–Majiang fault zone is nearly N-S-trending, measuring more than 200 km long, and corresponds to a N-S-trending gravity gradient zone. This fault zone was formed in the Indosinian period and controlled the generation of the Early Jurassic fault basins and early Yanshanian granite masses, and it was active in the Neoid period. The Fuchuan–Xiwan fault zone is more than 120 km long and consists of a series of parallel faults, and the Carboniferous system is thrusted above the Jurassic system. The fault was formed in the Indosinian period and remained active in the Yanshan period, controlling the distribution of the early Jurassic coal-bearing basins in the Fuchuan Xiaotian and Hexian Xiwan regions.

The eastern subzone is the Daoxian–Gupo Mountain fault depression zone, spreading at 111°40'E–111°55'E, with an intermittent length of 280 km and a width of 20–30 km, is composed mainly of the nearly N-S-trending Upper Paleozoic compound synclines and parallel fault zones. The northern section in the Shuangpai–Daoxian region includes the Zijingshan compound anticlines and the Shuangpai (Danjiang) compound synclines. The southern section in the Daoxian–Guposhan region features many "trough" folds. The Yanshanian Guposhan granites are transacted by the Baimangyuan–Hejie faults, and the Jurassic or Cretaceous system is also transacted by some N-S-trending faults. There are groups of quartz porphyry and granite produced along the faults. There was also seismic activity in the Neoid period.

(3) Leiyang–Linwu Fold-Fault Zone
The zone is spread at approximately 112°25'E–113°E, with a length of 160 km and a width of 60–80 km. It is generally composed of Upper Paleozoic asymmetrical N-S-trending compound syncline zones, and the axis is approximately located in the Changning Shuikoushan–East Linwu region, with tight and strong folds and faults developed in the same direction, forming the following sparsely and densely distributed fault depression zones.

(a) *Yuanjiayi–Xianghualing fault depression zone*: The folds are arranged sparsely at large spacing, and the local regions overlap unconformably by the Cretaceous system.

(b) *Shuikoushan–Linwu fault depression zone*: This zone is composed of a series of Paleozoic nearly N-S-trending linear folds. The faults are relatively developed, and most of them are concentrated in the convergent and turning positions or on both sides of the anticline of the tectonic zone, and some small Early Yanshanian acid intrusions and polymetallic deposits are distributed along the faults.

(c) *Taoshui–Yizhang fault depression zone*: It is exposed at approximately 113°E, and the main body is an Upper Paleozoic depression zone. Due to the influence of the N-E-trending system, its continuity is poor, and it features arc-shaped curves and northerly wide and southerly narrow

S-shaped folds. The nearly N-S-trending faults appear mostly on both sides of the anticline, and N-E-trending and N-W-trending conjugate shear faults are developed.[10] A large number of intermediate-acidic rock masses and basic dyke groups are distributed along both sides of the main fault and in the composite part of the N-N-E-trending system, and a very large Shizhuyuan tungsten–tin–bismuth–molybdenum deposit has formed in the contact zone of the Qianlishan granite masses.

The Hunan–Guangxi N-S-trending tectonic system might have been produced in the Caledonian period, formed in the Hercynian period, and finalized in the Indosinian period. It was obviously active in the Yanshanian period and was active in some regions from the Himalayan period to the Neoid period. It has obvious control over the Late Paleozoic (Carboniferous and Permian) deposition facies and coal and other deposition minerals, as well as the distribution of the Indosinian Zhonghuashan–Wutuan Granite zone, the early Yanshanian N-S-trending basic and ultrabasic rocks in southern Hunan, and the Jurassic and Cretaceous N-S-trending small basins. The compound relationship between the Hunan–Guangxi N-S-trending tectonic system and other tectonic systems is also obvious. For example, the N-S-trending tectonic system crosses the E-W-trending tectonic zones, forming a very obvious cross-fold zone. Among the folds, one type is T-shaped half-crossing folds, i.e., the nearly N-S-trending nose-like folds on the northern margin of the Jiuyishan, Dayaoshan and other large E-W-trending compound uplifts, which are manifestations of the N-S-trending double folding of the E-W-trending tectonic zone. The other type is "cross"-shaped full-crossing folds, i.e., the Yuankou and Liangchahe N-S-trending anticlines are superimposed with the Huashan and Guposhan E-W-trending compound anticlines, and there are Huashan and Gupo Mountain composite rock masses produced in the core of the superimposed section (Fig. 3.4). In the superimposed section of the N-S-trending tectonic system and the N-E-trending and E-W-trending tectonic systems, small Yanshanian granite masses and related non-ferrous metal deposits are often produced.

3.6.4 *East Anhui N-S-Trending Tectonic System*

There are a series of N-S-trending squeezing and torsional tectonic traces in the adjacent regions of Hebei, Shandong, Jiangsu, Anhui, Jiangxi and

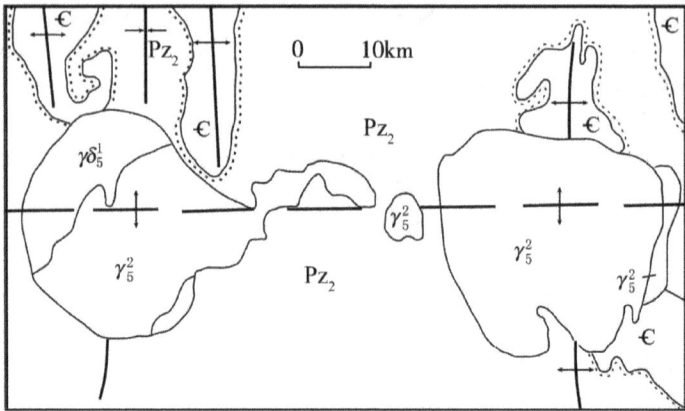

Fig. 3.4.   Layout of the Huashan and Guposhan masses produced along the superimposed uplift axis.

Fujian in eastern China, and the main body is located at 117°E–119°E. They have obvious zoning features and are approximately equidistant in the transverse direction. A more obvious concentration zone appears at a spacing of approximately 20'–30'. In the vertical direction, there are traces in the bedrock in the mountainous region from eastern Hebei to Fujian and Jiangxi, and their trend is generally 5°–10° east by north or west by north, forming N-S-trending tectonic zones with lengths of thousands of kilometers. In this system, there are not only relatively large compressional and torsional fault zones and fold zones but also large N-S-trending zone-like uplifts and depressions, as well as N-S-trending migmatite zones and granite zones. Gravity and magnetic data have shown that some of their components have good continuity and constitute tectonic zones with greater influence depths. Judging from the formation and evolution history, the northern, central and southern sections are not exactly identical or even very different, but rather, share some basic features from the Mesozoic to the Neoid period.

The Shouning–Lianjiang fault zone starts from Shouning in the north and ends at Guling in the south, with a length of approximately 180 km and a width of 3–10 km. It is composed mainly of high-angle thrust faults and N-S-trending granite masses. At the end of the Late Jurassic, a series of thrust faults formed in the volcanic rocks, accompanied by alteration zones, schistosity zones, etc. The Cretaceous monzonitic granites intruded along the fault zones, forming a tectonic magmatic rock zone with a length of 70 km and a width of 10–25 km. After the Cretaceous, N-S-trending compressional faults appeared and cut the late Yanshanian granite masses.

The East Anhui N-S-trending tectonic system is large in scale and spans 3 longitudes, with a length of approximately 2,000 km. Its tectonic traces feature strict orientation, approximately equal width and distance, poor continuity, compound natures and different development processes. It resulted mostly from squeezing and torsion to the left in the early period and squeezing and torsion to the right in the late period. It first appeared in the early Cambrian strata in the North China block, Qianxi Group and Wuhe Group with the main body at approximately 118°E and the Zhangbaling Group metavolcanic rocks in the Yangtze block and was equivalent to the products in the Jinning period. In contrast, the Caledonian deformation, metamorphism and migmatization zones are found in the Yangtze block and the Cathaysia block and contain paleotectonic zones.[11] They penetrated from north to south only in the middle and late periods of the Yanshanian, forming the current east-north Anhui N-S-trending tectonic system, and they were active to some degree in the Neoid period. The general appearance and system attribution in the pre-Yanshanian period needs to be further studied. In general, it crosses through the three major E-W-trending tectonic systems in eastern China from north to south. Its early tectonic zone has a certain restrictive effect on the formation and development of other tectonic systems in the region. For example, the Tanlu fault zone of the North Jiashan N-N-E-trending system is distributed in a nearly N-S-trending direction. Due to the multiple activities of the E-W-trending system, the mechanical properties of the N-S-trending zone changed mostly to tension and torsion in the E-W-trending region. Conversely, the Neoid activities of the southeast Anhui N-S-trending tectonic system were also inevitable to transform other tectonic components in the region. For example, the N-E-trending and N-N-E-trending fault tectonic zones are often used or merged by a group of torsion fault surfaces. Consequently, the left compression and torsion turned into right compression and torsion. A stepped tectonic combination of N-N-E-trending tension and torsion faults was encountered by PetroChina in the North China Plain. Therefore, the control effect of the East Anhui N-S-trending tectonic system on the migration and accumulation of oil and gas in the region cannot be ignored.

3.6.5 *Taiwan Coast N-S-Trending Tectonic System (Chinese Taiwan–Philippines N-S-trending Tectonic System)*

The Taiwan Coast N-S-trending tectonic system can be approximately divided into three zones, namely, the central Taichung–Pingtung fold-fault zone, the eastern Lvdao–Lanyu N-S-trending volcanic eruption zone and the western Penghu Waterway N-S-trending depression zone. They are connected to the Philippine Luzon zone to the south, forming the Chinese Taiwan–Philippines N-S-trending tectonic zone.

(1) Lvdao–Lanyu Volcanic Rock Zone

The zone is located at 121°10'E–123°E and is composed of N-S-trending ridges and volcanic islands above the water's surface. It is connected to the coastal mountains to the north and to the Batan Islands, the Pabuyan Islands and Luzon to the south. The southern section has a width of approximately 185 km, and the northern section has a width of 26 km at Lanyu. Both sides are bounded by large high-angle thrust faults, forming a very striking N-S-trending volcanic clastic uplift zone on the seabed. In China, this zone has a length of approximately 200 km and a width of 20–40 km and is composed mainly of Miocene–Pliocene andesite and andesitic volcanic clastic rocks. Most of the volcanic rocks belong to the porphyry basalt series, while the Lvdao island features the high-aluminum basalt series, similar to the Chimei magmatic complexes exposed in the middle section of the coastal mountains, indicating that they belong to the same volcanic island arc zone. This zone has obvious indications on the gravity and magnetic maps as follows. The isogravity line is N-S-trending, and it is a high gravity value region. The isogravity line is also N-S-trending and is a magnetic anomaly zone on the magnetic map, which is clearly separated from the smoother magnetic band on the west. The volcanoes were active from north to south in the Luzon arc zone; that is, the ages of volcanic rocks show a trend from old to new in the Chimei magmatic complexes in the Coastal Mountains from Lanyu, the Batan Islands and the Babuyan Islands to Luzon. No active volcano is found in Chinese Taiwan's volcanic island arc zone, but there are some volcanoes in the Batan Islands, Babuyan Islands and Luzon.

(2) Coastal Mountains Fold Zone

The zone is adjacent to the Pacific Ocean on the east and is bounded by the Taitung Rift Valley on the west. It is approximately 135 km long and has a maximum width of approximately 10 km on the Chinese Taiwan.

Its southern section is the N-S-trending "Philippines Active Zone" composed of Luzon, Zhushayan–Mindanao, etc. It is an integral part of the island arc zone around the Pacific Ocean. The main tectonic lines in its interior and on its margins extend in a N-S-trending direction, with wave-like upward curves, forming a series of eastward or westward convex arc zones dominated by faults and folds. The Miocene and Pliocene systems are subject to various degrees of deformation and transformation, and the Mesozoic and Paleogene strata are subject to strong folds and thrusts, forming some compound anticlines and rift troughs. The intermediate-basic magmatic activities were relatively strong, and the southern Madrid section was covered by Quaternary volcanic rocks. Quaternary volcanic rocks are developed in the central-southern section. The famous reverse S-shaped Philippine fault zone is interspersed in the entire active zone, forming eastern or western arc-shaped tectonics, i.e., the Sama Arc. This tectonic zone still clearly spreads in the Boni Bay region of the Sulawesi Island, and its eastern section may include the Philippine Trench and both banks of the Maluku Strait,[12] forming a N-S-trending tectonic system with a length of approximately 3,000 km, which is a modern highly active tectonic system. The northern extension of this system is still unknown, and the relationship between it and the N-S-trending compression and compression-torsional tectonic zones is intermittently distributed along the eastern coast of Jiangsu–Zhejiang, the eastern Jiaodong Peninsula. The northern region at the same longitude has not yet been studied.

3.6.6 *Mudanjiang N-S-trending Tectonic System*

The Mudanjiang N-S-trending tectonic system is exposed mainly in the eastern part of Heilongjiang Province, so it is also called the East Heilongjiang N-S-trending tectonic system. It extends northward to the west of the Bureya Range in Russia and southward to the northeast of the Korean Peninsula. Its main components spread at approximately 128°E–132°E and include the Jiamusi–Laoyeling paleo-uplift zone and the Paleozoic depression zones on the eastern and western sides, as well as several major fault zones (Fig. 3.5).

(1) Jiamusi Uplift Zone
This zone, also known as the Laoyeling Middle Uplift Zone, is located at approximately 129°E–131°E and is a N-S-trending zone, passing through Heilongjiang and extending into Russia to the north and through Jiayin,

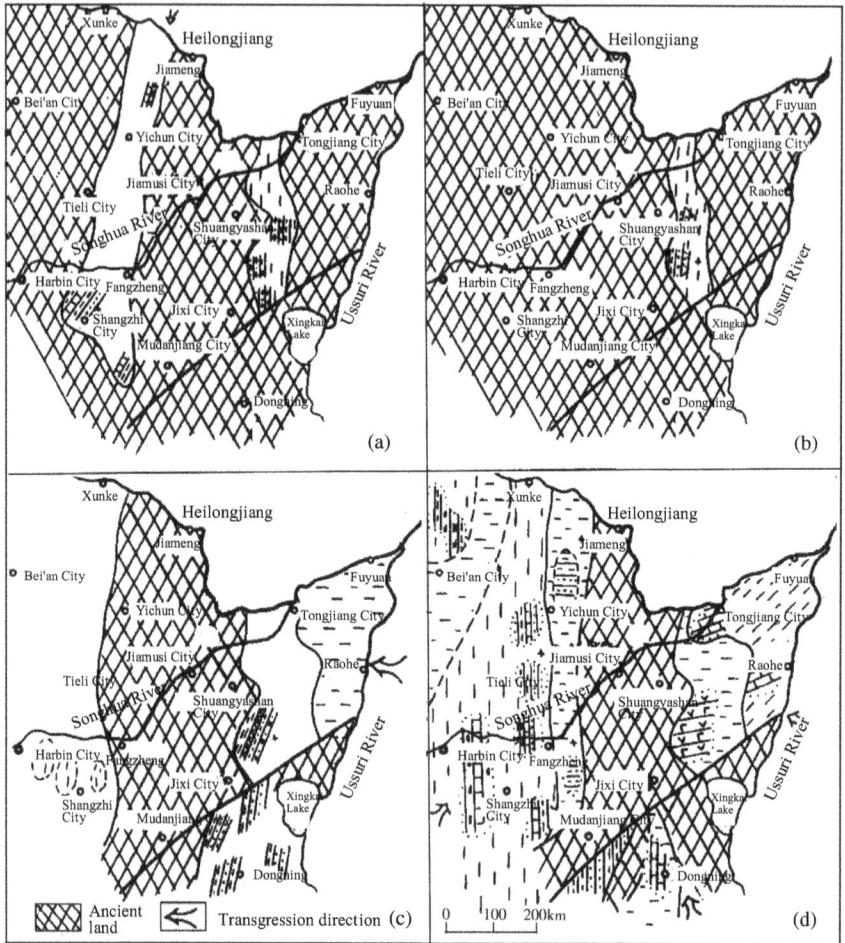

Fig. 3.5.   Schematic diagram of Mudanjiang N-S-trending zone evolution in the Late Paleozoic.

Hegang, Jiamusi and Muling–Hailin to the south, and is intersected obliquely by the Dunhua–Mishan N-E-trending fault zone and Changting N-E-trending fault zone. The middle section is intersected by the Yilan–Yitong fault zone and the Taxi–Linkou N-W-trending fault zone. It includes the western Zhangguangcailing marginal uplift zone, with a width of approximately 170 km and a length of more than 700 km. The Meso-New Proterozoic Heilongjiang Group green schist and carbonate systems exposed in the zone are products of a deep-sea closed environment. In the Mashan region, the paleo-continent core is composed of the Neoarchean Mashan Group, and the Paleoproterozoic system is a continental marginal active formation. In addition, the Lvliang Movement (Xingdong Movement) folds and uplifts form N-S-trending short-axis anticlines and synclines near the paleo-continental core, and the destroyed paleo-continental core is embedded by *ex-situ* granodiorite inlays in the Paleoproterozoic nearly N-S-trending fold zone. The uplifts on the Mesoproterozoic continental crust led to the tension of the Jinning continental crust, forming the Luobei–Taipinggou, Yilan, and Mudanjiang N-S-trending rift troughs. In the early Jinning Movement (approximately 1000 Ma), the Heilongjiang Group metamorphic rock system (922 Ma, but most of the isotopic age is reflected in the Phanerozoic) was formed at the rift axis, the N-E-trending fold zone was welded with the paleo-continental mass by the granodiorite formed during the same tectonic period, and the muscovite, epidote and granite formed at the end of the tectonic period, creating a composite paleo-tectonic pattern. The high green schists-low amphibolite facies metamorphic rocks of the Majiajie Group formed in local regions in the Late Yaning Period, and the late Jinning Movement (approximately 800–850 Ma) folds and uplifts were intruded by muscovite, tourmaline and granite. They are developed into active troughs west of the Jiamusi uplift zone (west of the Mudanjiang fault) in the Sinian (Zhangguangcailing cycle), forming the Zhangguangcailing Group and the Huangsong Group. The fold uplifts at the end of the Sinian became a marginal uplift zone on the western margin of the Jiamusi block, and the southeastern section became a Taipingling N-E-trending tectonic zone. In the early Caledonian, the uplifts were dominant; only the eastern Luobei and Xingkai Lake regions subsided locally to form the Lower Cambrian caprocks. The Middle Cambrian-Silurian system was missing, and the uplift zone was still in the uplift and denudation stage in the Hercynian. A large number of granites were emplaced during the Late Indosinian Movement, forming a powerful Indosinian granite zone. In the Yanshanian period, the Hegang and Jiayin depressions formed coal-bearing basins.

(2) Baoqing-Mishan N-S-Trending Fold-Fault Zone
The zone is located east of the Jiamusi uplift zone, spreading in a N-S-trending arc shape and slightly protruding to the west. It is covered by the Sanjiang Basin to the north and is adjacent to the Wandashan active tectonic zone to the east. It is a transition zone between the uplift zone and

the active zone and formed mainly in the Late Paleozoic. The Late Indosinian granite zone was formed in the Yichun–Yanshou region and is superimposed on the Caledonian tectonic magma zone.[13] It spreads to Jilin in the south and to Russia in the north, forming a tectonic magma zone with a width of approximately 150–200 km and a length of more than 1,000 km. Some sections are also sporadically distributed along the Shuangyashan and Mishan regions southeast of the Jiamusi paleo-uplift zone.[14] The metamorphism of green schist facies and the intrusive activities of basic rocks are related to the deformation and intermediate-acid magmatism in the Late Indosinian movement.

The above discussion shows that the main body of the N-S-trending tectonic system formed in the Caledonian period and was finalized in the Indosinian period. It merged or contained the pre-Caledonian N-S-trending tectonic magmatic zone.

References

1. Feng Fuguo *et al. Natural Gas Geology in China.* Beijing: Geological Publishing House, 1995.
2. Zhu Xia. *Mesozoic and Cenozoic Basin Structure and Evolution in China.* Beijing: Science Press, 1983.
3. Ren Jishun. *Tectonics and Evolution of China.* Beijing: Science Press, 1980.
4. Geological Mechanics Research Mapping Group, Academy of Sciences. *Specification of Tectonic System Map of the People's Republic of China (1: 4,000,000).* Beijing: Geological Publishing House. 1978.
5. Liu Guangding *et al. Geological and Geophysical Characteristics of China's Sea Areas and Adjacent Areas.* Beijing: Science Press, 1992.
6. Liu Baoyao. *Crustal Evolution and Mineralization of Ancient Continent in South China.* Beijing: Science Press. 1993.
7. Zhai Guangming *et al. China Petroleum Geology (Volumes 13 and 14).* Beijing: Petroleum Industry Press, 1996.
8. Dai Jinxing, Wang Tingbin. *Formation Conditions and Distribution Laws of Large and Medium-sized Natural Gas Fields in China.* Beijing: Geological Publishing House, 1997.
9. Xiao Xuchang *et al. Geotectonics in the North of Xinjiang and Adjacent Areas.* Beijing: Geological Publishing House, 1992.
10. Zhang Yuchang *et al. Prototype Analysis of Petroliferous Basins in China.* Nanjing: Nanjing University Press, 1997.
11. Zhang Fuli *et al. Natural Gas Geology in Ordos Basin.* Beijing: Geological Publishing House, 1994.

12. Yuzhu Kang. *Characteristics of Paleozoic Marine Oil Generation in China.* Urumchi: Xinjiang Science and Technology Health Publishing House, 1995.
13. Yuzhu Kang. *Main Structural Systems and Oil and Gas Distribution in China.* Urumchi: Xinjiang Science and Technology Health Publishing House, 1999.
14. Yuzhu Kang *et al. Paleozoic Marine Oil and Gas Field in Tarim Basin.* Wuhan: China University of Geosciences Press, 1992.
15. Li Siguang. *Introduction to Geological Mechanics.* Beijing: Sciences Press, 1973.

Chapter 4

N-E-Trending Tectonic Systems

Yuzhu Kang, Zongxiu Wang, and Huijun Li

Abstract

In this chapter, N-E-trending tectonic systems are introduced, including those in China, New Zealand in the Pacific, northwestern Europe, the eastern United States, eastern South America and eastern Africa.

Keywords: Tectonic System; N-E-trending; Type

4.1 N-E-Trending Tectonic System in China

This tectonic system is a ξ-shaped tectonic system composed of N-E-trending and S-W-trending fold zones, which formed during the Caledonian and Indosinian Movements on the eastern Chinese mainland. It is a deformation and metamorphic zone formed by strong shear and distortion of pre-Late Triassic strata and rock masses. It has mostly been unconformably covered by the strata since the Late Triassic and generally spreads at 45° north to east. The system is dominated by N-E-trending fold zones (or uplifts and troughs), accompanied by compression and compression-torsion faults, as well as large-scale N-E-trending metamorphic zones, granite zones and volcanic rock zones in some regions. The N-E-trending tectonic system developed mainly from the Paleozoic to the Middle Triassic. It is an earlier ξ-shaped tectonic system in the Cathaysia system, which was generally subject to deformation and metamorphism in the Caledonian Movement and the main act of the Indosinian Movement. It finalized from the late Middle Triassic to the early Late Triassic, playing an important role in controlling the Late Paleozoic and Triassic deposits and magmatism. It featured mainly plastic deformation, accompanied by low-temperature and medium and high-pressure dynamic deformation and metamorphic zones generally at 45°–50° north to east. Due to the intersection and partition of the E-W-trending tectonic zones and differences in the basement structure of different regions, the development degrees and the deformation features differ across sections, as described in the following sections.

4.1.1 N-E-Trending Tectonic System in Northeast China

The N-E-trending tectonic system in this region is relatively developed, with obvious traces and older ages. It appeared in the Proterozoic, was finalized in the Paleozoic to Late Triassic and was transformed by subsequent tectonic movements.

(1) Taipingling–Hulin Tectonic Zone
This zone spreads on both sides of the Dunhua–Mishan N-E-trending intermittent zone along the southeastern margin of Heilongjiang Province. It is composed of intermittent N-E-trending folds, compressional fault zones and Indosinian intermediate-acid magmatic rock zones, and extends southwest into Jilin Province. It is faintly evident that there was a N-E-trending tectonic magma deformation and metamorphic zone in the pre-Yanshanian period. The Taipingling–Hulin uplift zone extends southwest to the Wangqing region of Jilin, and the Hulin–Hutou region extends northeast. The Neoproterozoic–Paleozoic–Triassic systems were intermittently exposed, and the Archean crystalline basement was also exposed in the Hulin region, forming a compound anticline fold zone with the Neoproterozoic metamorphic rock system as the core, the Paleozoic and Triassic systems as the wings and the fold axis at 40°–50° north to east.[1] In the Sino-Russian border region northeast of Taipingling, the Permian–Triassic system unconformably covers the Neoproterozoic metamorphic rock system in the N-E-trending direction. In the region near Dongning, there is a Carboniferous-Triassic fold tectonic zone, which is unconformable from the Upper Triassic system, and the lower strata spread in the N-E-trending direction, superimposed by the N-N-E-trending and nearly N-S-trending tectonics in the later period. The axis of the Taipingling

compound anticline is between the eastern side of Taipingling and Suifenhe city, accompanied by the developed N-E-trending compression and compression-torsional faults with varying scales, including the Suiyang–Huangnihe faults, the East Taipingling slope faults, Dongning faults, Changting faults, etc. They formed in the Paleozoic and at the end of the Triassic and were mostly intruded by Yanshanian rock masses without intersection with Yanshanian rock masses. The Dunhua–Mishan fault zone has been active for a long time, and its mechanical properties have undergone several obvious changes. It has been an active fault zone in multiple periods. It was a torsion fault zone in the late Jinning period. In the Caledonian period, it was transformed into a component of the Taipingling N-E-trending tectonic zone. A Permian–Triassic system unconformably covered the Sinian Yanwangdian Formation, Yangmu Formation, etc., and the nature of the faults changed to compression and counterclockwise torsion. In the Indosinian Movement, after strong compression and counterclockwise torsion, the magmatic rock zone spread along its southern side, forming a strong N-E-trending active magmatic zone, which is an important part of the Cathaysia tectonic system in northeastern China.

(2) Zhangguangcailing–Laoyeling Tectonic Zone

This zone spreads between the Dunhua–Mishan fault zone and the Yilan–Yitong fault zone, with its main body distributed along the Zhangguangcai Ridge at the junction of Jilin and Heilongjiang to the southwest, Songhua Lake on the western side of the Jiao River, the Laoyeling and Yongji–Panshi regions and its axis being a Late Hercynian–Indosinian granite zone. The Upper Paleozoic (mainly Permian) and Triassic semi-surrounding magmatic rock zones are distributed on the western and northwestern sides of the rock zone. Isotope age test results indicate that the Zhangguang Cailing granite zone formed in the Indosinian. The isotope age of the Laoyeling granite zone in Jilin was mostly 210–251 Ma, followed by 290 Ma and 180–220 Ma, so they are mainly the products of Late Hercynian–Indosinian intermediate-acidic and acidic magma intrusions. The strata preserved today constitute a compound syncline with the Triassic system as the core, which is known as the Jilin N-E-trending compound syncline in the "Regional Geology of Jilin Province". Therefore, this magma deformation tectonic zone is a product of the late N-E-trending system, which is compounded with the Jiamusi uplift zone in the northeast and compounded with the Yinshan–Tianshan E-W-trending zone in the southwest.

(3) Hunjiang Fold-Fault Zone

It is a N-E-trending fold-fault zone formed in the Late Hercynian–Indosinian Movement on the basis of the Hunjiang Depression from the Qingbaikou period to the end of the Paleozoic. Its main body is a compound syncline, its core is a Carboniferous-Permian system, and its two wings are composed of Ordovician and Qingbaikou systems. However, there was no strong tectonic deformation and metamorphism between the Qingbaikou system and the Permian system, but there were some discontinuities between the end of the Ordovician and the beginning of the Carboniferous. The formation features and biota in the Qingbaikou to Sinian systems can be compared in groups with those in the Jangsu Huaibei and East Liaoning regions. This zone extends southward, passes through the Liaodong Peninsula to the western side of the Jiaodong Peninsula, and is integrated with the Xuzhou region, forming a group of Neoproterozoic-Paleozoic ξ-shaped depression zones at 40°–50° north to east. The deposits and formations have transitional features in South China and North China but are closer to those in South China.[2] Under the N-W-trending and S-E-trending compression stresses in the Indosinian Movement, a strong deformation zone formed, and the lower Jurassic and Middle–Upper Jurassic systems were extensively unconformable on the underlying strata, accompanied by a large amount of compression and compression-torsional fault zones, so this zone is known as the Indosinian fold-fault zone. The Indosinian fold-fault zone is generally a long and narrow zone at 40°–50° north to east, and its axis has a small intersection angle with the axis of the pre-Qingbaikou system fold-fault zone, without metamorphism in the rock formations and Indosinian magma intrusion zone.

Parallel to the Hunjiang fold-fault zone, the Yangzishao fold-fault zone spreads on the northwestern side of the Longgang compound anticline. It has the same features and development history as the Hunjiang fold-fault zone but a smaller width, clear traces and a stable direction. Its southwestern section is unconformably covered by the Upper Jurassic system, and it is a depression fold zone of the N-E-trending system in this region.

The main fault zones associated with the above two fold-fault zones are the Sanyuanpu–Yangzishao fault zone and the Hunjiang–Wangou fault zone, which are crustal faults controlling the Neoproterozoic–Paleozoic deposits and Mesozoic inhibitory activities. They show a clear geophysical prospecting response, good surface presentation, an orientation at approximately 50° north to east, significant compression features, and they pass diagonally through the Liaoning and Jilin regions. These two fault zones actually control the boundary between the Hunjiang fault depression zone and the Yangzishao fault depression trough. The Hunjiang–Wangou fault zone extends into Liaoning and is called the Huanren–Zhuanghe fault zone.

(4) Yalu River–Suifenhe Fold-Fault Zone

The main body of the fold-fault zone is the Suifenhe–Yanbian compound syncline and the Yalujiang–Song River fault zone. This fold-fault zone spreads on the Taipingling compound anticline and the southeastern side of the Hunjiang fold-fault zone and extends into Russia on the northeast. From the formation data, the product in the Wandashan region is integrated with that in the Yanbian region in Heilongjiang Province, but the former is located on the northern side of the Dunmi fault zone. In terms of fault distribution, the Yalujiang-Song River fault zone extends south of Xingkai Lake. The middle section of the fold-fault zone is located in the Yanbian region of Jilin Province, forming a Yanbian N-E-trending compound syncline zone. The Lower Paleozoic active deposition formations are exposed sporadically on the margin, and the middle part is covered by new strata. However, the Carboniferous system is dominated by marine carbonate rocks and supplemented by terrigenous clastic rock formation. The lower Permian system is composed of terrigenous clastic rocks, volcanic rocks, and carbonate rocks, and the upper Permian system is composed of N-E-trending intermediate-acid volcanic rocks and piedmont molasses, on which Cathaysian flora such as large feather ferns are grown. The Triassic system is dominated by volcanic rocks with sandy slates and thin coal seams, distributed in a N-E-trending direction, on which the Jurassic–Cretaceous continental basins are unconformably covered.

The Yalu River–Song River fault zone was formerly known as the Yalu River deep fault (large fault). The fault zone extends 45°–50° north to east from Liaoning along the Yalu River, Ji'an in the Tonghua region of Jilin, and the Song River and Suifenhe region of Heilongjiang, with a length of more than 1,000 km. The fault zone is clearly shown on the aeromagnetic map, and gravity data have shown that there is a N-E-trending fault zone with good continuity under the Linjiang–Song River large basalts.[3] The formation data indicate that the fault zone formed in the Hercynian–Indosinian period, the southwestern section of the fault zone is located on the southeastern side of the Hunjiang trough formed in the

Early Indosinian period, and the northern section controls the Late Haitian–Early Indosinian basic and ultrabasic rock masses. Before the Yanshanian stage, the fault zone showed anti-clockwise compression and shear, with a shift distance of approximately 30 km, which turned into extensional or extensional shear activities in the Late Triassic–Jurassic period, controlling deposition and volcanic eruption activities in the Late Triassic, Late Jurassic and Cretaceous.

4.1.2 N-E-Trending Tectonic System in North China

The North China block has been a stable environment for a long time since the Lvliang Movement in the late Paleoproterozoic, and there was no obvious unconformable interface from the Changchengian period to the Triassic period. Moreover, there was only uneven regional uplift and subsidence in the Jinning Movement and Caledonian Movement with broad effects on the evolution and development of China's continental crust, resulting in the Sinian and Silurian–Devonian systems being missing widely in the North China block. The Cambrian and middle Carboniferous systems were generally in parallel unconformable or overlying unconformable contact with the lower Meso-New Proterozoic or Lower Paleozoic rock systems. The deformation and metamorphic zone formed in the Lvliang Movement and the rift zone formed at the beginning of the Changchengian period show N-E-trending ancient tectonic traces and masses in the Shanxi continental platform and the Yanshan–Yinshan region. Therefore, whether there is a N-E-trending tectonic system in the North China block, especially the Shanxi continental platform, has long remained unknown.

(1) N-E-Trending Tectonic Zone in Shanxi
Changes in the thickness of the Cambrian-Ordovician groups in the Shanxi continental platform indicate that the early Paleozoic tectonic zones were N-E-trending and N-W-trending depressions and uplift zones opposed to each other (Fig. 4.1). In the Late Paleozoic, the Shanxi continental platform was mainly near-N-S-trending and near-E-W-trending uplift zones, as well as some N-E-trending trough basins, i.e., the Wutai–Fenyang trough in the Taiyuan period and the Xiaxian–Qinshui uplift in the Shanxi period. In the Late Carboniferous period, there were only the N-W-trending Qinshui Submarine Uplift and the Zhongtiao Paleo Uplift

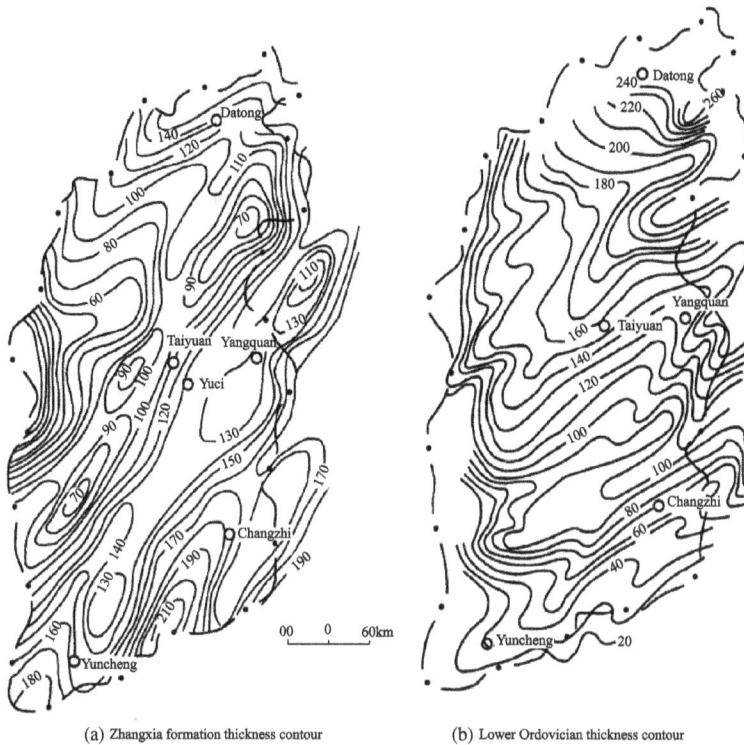

(a) Zhangxia formation thickness contour   (b) Lower Ordovician thickness contour

Fig. 4.1.   Schematic diagram of Paleozoic tectonic evolution in Shanxi, China. (a) Contour map of the Zhangxia formation thickness and (b) contour map of lower ordovician thickness.

on the Shanxi continental platform in the Shanxi period, showing the inherited evolution of the N-E-trending tectonic zones. In the Permian Lower Shihezi stage, there was only the N-E-trending Jincheng Depression east of Changzhi.

In addition, from the distribution of the Late Carboniferous Tingidae fossil zones on the North China block, there are still indications of the N-E-trending tectonic zone in this period (Fig. 4.2).

The above data indicate that there are Paleozoic N-E-trending tectonic traces on the North China block, indicating the manifestation of the N-E-trending system on the stable North China block. It features mainly inherited activities of broad and gentle uplifts and faults, without obvious tectonic deformation.

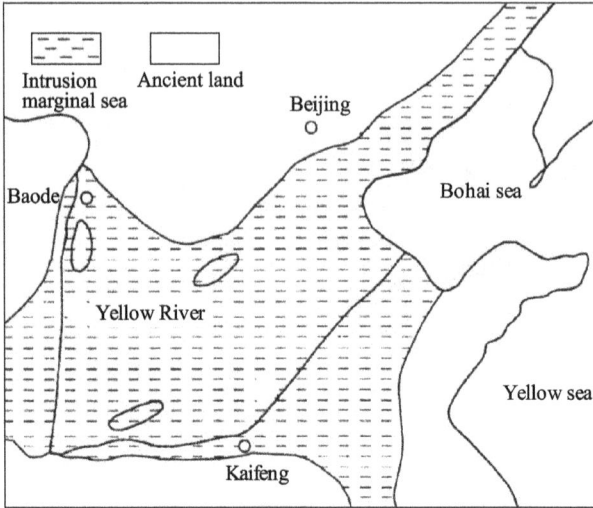

Fig. 4.2.   Paleogeography of the North China region in the Late Carboniferous.

(2) N-E-Trending Tectonic Zone on the Northern Margin

More than 10 years of field investigations, paleotectonic screening, deposition lithofacies paleogeography and paleotectonic analysis from Li Jinrong and others has shown that the Yinshan–Yanshan–Liaodong region along the northern margin of the North China block has a long-term development history and was finalized in the Indochina period. It is mainly a set of N-E-trending fold zones, thrust zones and tectonic magmatic zones. It includes the North Taihang Mountains–Jundushan fold zone, the Xinglong Lishugou–Wangxiaogou fold zone, the Tangshan–Lingyuan fold-fault zone, the Liugezhuang–Chaoyang fold-fault zone, the Sandaogou–Fuxin fold-fault zone and the Jinzhou–Yiwulushan uplift fault zone,[4] some of which are discussed next.

(a) *North Taihang Mountain–Jundushan fold zone*: This fold zone is divided into three sections and is a ξ-shaped oblique array at 40°–50° north to east, with the northern section reversely compounded or transected by the E-W-trending fault zones and the southern section composed of the Meso-New Proterozoic–Paleozoic Baijian compound anticline, Dongxinghe syncline, Chashan anticline, etc. The middle section is composed of Qijiazhuang folds, Shijingshan folds, Fenghuangshan syncline, Shimenying anticline, Wanfosi–Shijingshan syncline, Sangyucun–Gaojing anticline, Xiangshan syncline, Muchengjian syncline, etc. These single anticlines and synclines are composed of shallow metamorphic Paleozoic and Lower Triassic systems, which the Late Triassic or Early Jurassic basins cover unconformably.

(b) *Tangshan–Lingyuan fold-fault zone*: The southern section of this zone is the Tangshan fold zone distributed on the North China Plain, which is composed of the Tangshan syncline and Gaoluogu syncline with Carboniferous–Permian coal-bearing strata as its core. The northern section is the Lingyuan N-E-trending fold-fault zone distributed from Xifengkou to Xiangyang, with a length of 300 km, and shows a ξ-shaped oblique array. The Lingyuan Taipinggou–Daoerdeng folds are linear folds composed of the Sino-New Proterozoic–Triassic system, and its anticline axis is at 25°–30° north to east with a length of 26 km. A group of faults parallel to the fold axis contains the diorite emplacement with an age value of 204 Ma. In the Yushulinzi region, faults and magmatic activity are dominant. For example, the Yebaishou fault zone is distributed at 60° north to east and extends in a gentle wave shape that is 30 km long. The zone is sandwiched with lenticular rock slices or accompanied by secondary closed folds and thrusts. In addition, there are intrusions of Indosinian granite at an age of 224 Ma and monzonite at an age of 228 Ma along the faults.

(c) *Liugezhuang–Chaoyang fold-fault zone*: This zone can be divided into three sections. The southern section includes the Magezhuang syncline, the Sanlinggao anticline and the Sanchakou syncline composed of Sino-New Proterozoic short axis folds. The middle section is located west of Jianchang and is a Nanyao–Laochangzi inverted syncline with a length of 39 km.[5] The northern Gongyingzi section is composed of the Wumishan–Permian system at 20° north to east, the southwestern Gongyingzi section is composed of the Gaoyuzhuang–Triassic system at 50° north to east and the Kazuo–Yangshugou syncline has a complete shape.

(3) N-E-Trending Tectonic Zones in the Ordos Basin
A set of N-E-trending fault zones and uplifted depressions are obviously manifested in the basement and Late Paleozoic system in this region. The basement fault zones from north to south are the Dingbian–Yulin fault zones and Wuqi–Suide fault zone and include the Huanglong and Pucheng fault zones (Fig. 4.3).

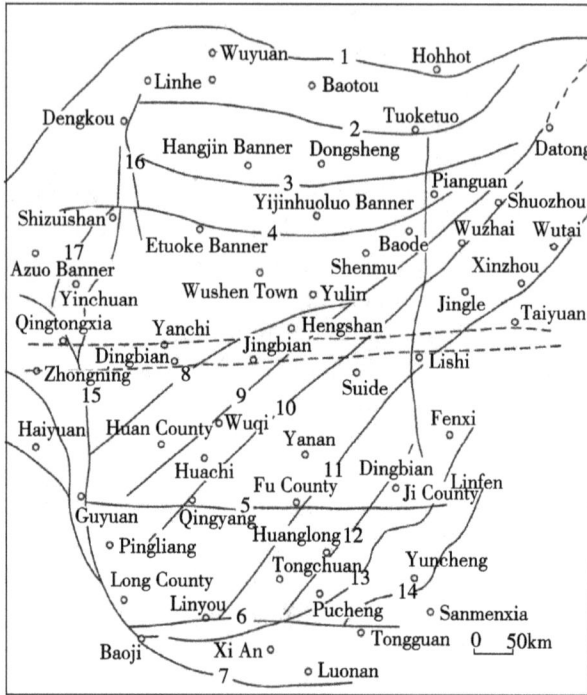

Fig. 4.3.   Distribution features of faults in the periphery and basement of the Ordos Basin.

Notes: E-W-trending fault zones: (1) Wuyuan North–Hohhot fault; (2) Linhe–Tuoketuo fault; (3) Hangjin Banner–Dongsheng fault; (4) Shizuishan–Pianguan fault; (5) Guyuan–Linfen fault; (6) Linyou–Tongguan fault; (7) Baoji–Luonan fault.

N-E-trending fault zones: (8) Dingbian–Yulin fault; (9) Huan County–Datong fault; (10) Qingyang–Shuozhou fault; (11) Fu County–Lishi fault; (12) Huanglong fault; (13) Pucheng fault; (14) Yuncheng fault.

N-S-trending fault zones: (15) Yinchuan-Guyuan fault; (16) Zhuozishan fault; (17) Yinchuan West fault.

(a) *Qingyang–Shuo county fault*: This fault extends in a N-E-trending direction, which is parallel to the Fu County–Lishi fault and is a fault in the Paleoproterozoic and Neoarchean systems. The reflection seismic data have shown that there is no fault in the deposition caprock, so they might have been active mainly in the Neoarchean–Paleoproterozoic period.

(b) *Fu County–Lishi fault*: This fault also extends in the N-E-trending direction, starting from Xinzhou in Shanxi in the north and passing through Lishi and Fu County to Yongshou in the south. The fault is clearly reflected on the aeromagnetic map, forming a boundary between the positive and negative magnetic fields. It is also reflected on the gravity anomaly map. It can be inferred that the basements on the northern and southern sides of the fault are the Neoarchean and Paleoproterozoic systems, respectively, so these systems also include faults that control the lithology of the basement and are supercrustal faults. The Fu County–Lishi fault was formed in the Paleoproterozoic, and its activity weakened after the Proterozoic. This fault forms the southern boundary of the Proterozoic Rift Valley (Shanxi–Shaanxi depression and tensional trough).

(c) *Basement aeromagnetic anomaly zone*: This anomaly zone includes two positive anomaly zones (Long County–Yan'an and Hancheng) and two negative anomaly zones (Wuqi–Yulin and Huangling–Lishi) (Fig. 4.4).

(d) *Basement gravity anomaly zone*: This anomaly zone consists of Huachi–Xing County, Huangling–Lishi and Hancheng–Linfen positive anomaly zones sandwiched with two negative anomaly zones.[6]

The late Paleozoic system is composed of the N-E-trending Luliang uplift and Yan'an depression. The N-E-trending Shenmu–Yichuan depression zone and the Wushenqi–Zhidan uplift zone occurred in the Cambrian. This system was still active in the late Paleozoic and controlled its deposition to a certain extent (Figs. 4.5 and 4.6).

4.1.3 *N-E-Trending Tectonic Systems in the Anhui, Shandong and Jiangsu Regions of East China*

This region is located on the southeastern margin of the North China block and near the Yangtze block. There have been no strong tectonic changes, but some giant N-E-trending uplifts and depressions appeared periodically in the Sinian-Middle Triassic period. The strong Indosinian Movement caused severe deformation in this region at the end of the Middle Triassic and the early period of the Late Triassic, forming a set of N-E-trending strong deformation and metamorphic zones, i.e., the N-E-trending tectonic system.[7] They are mitered and partially reconnected to the N-E-trending Paleo uplift zone at a slight intersection angle.

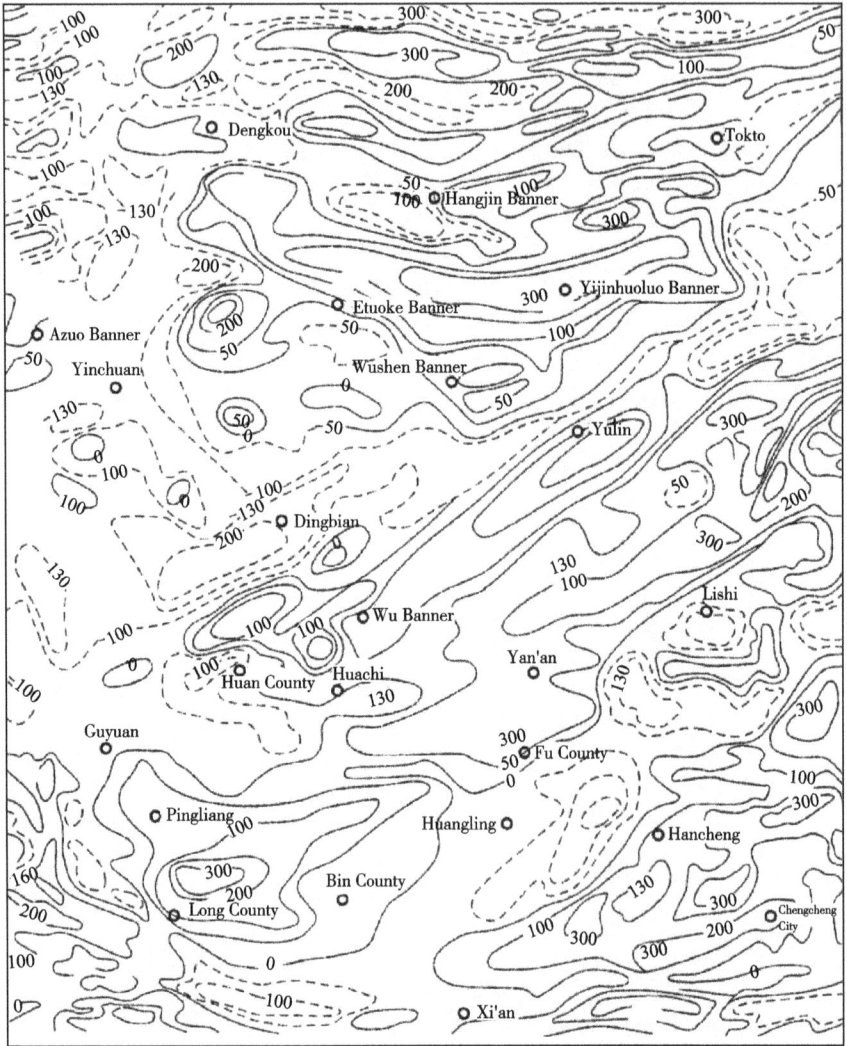

Fig. 4.4.   Aeromagnetic anomaly map of the Ordos Basin (Changqing Oilfield, 1983) (Unit: nT).

Except for some tectonic zones, they have quite different degrees of deformation and metamorphism. The N-E-trending tectonic system formed by the Indosinian Movement includes mainly the following tectonic zones in this region.

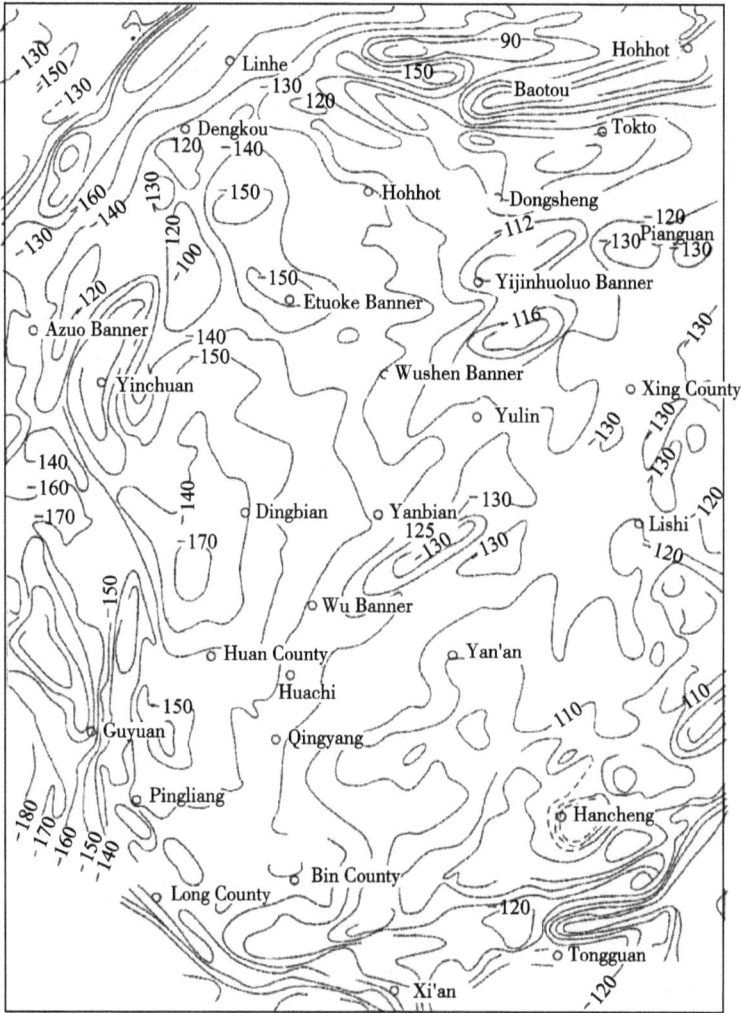

Fig. 4.5. Bouguer gravity anomaly map of the Ordos Basin (Changqing Oilfield, 1983) (Unit: mGal).

(1) Liaonan–Xuhuai Compound Syncline

This syncline spreads on the northwestern side of the Dabie (Huaiyang)–Jiaodong uplift zone, with the Sinian system as the main body, forming a N-E-trending S-shaped tectonic zone. Some predecessors carried out

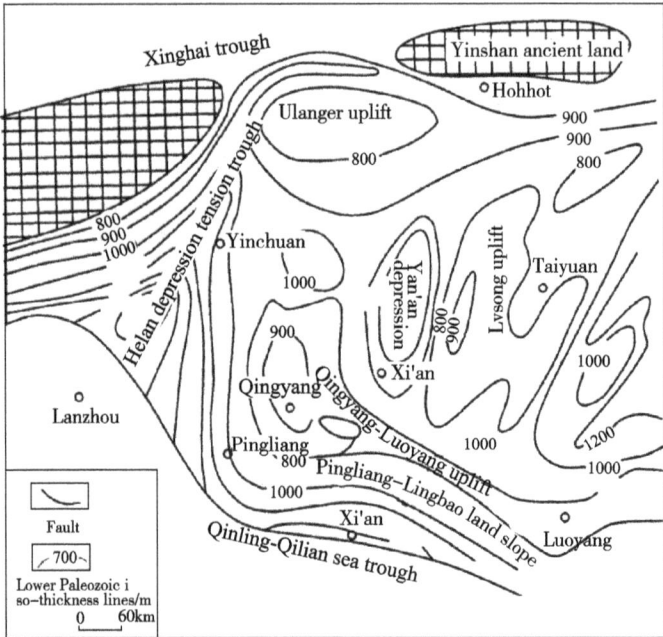

Fig. 4.6. Schematic diagram of the Early Paleozoic tectonics in the Ordos Basin and adjacent regions (modified).

many comparisons of the Sinian strata in the Xuhuai and Liaonan regions and affirmed that they belonged to products in the same environment and the same period; that is, it was an S-shaped trough that passed through central Shandong at that time and was dominated by limited platform facies and silica-containing calcium–magnesium carbonate deposits. This strata extends southwest to Sishili Mountain adjacent to the boundary between Henan and Anhui. The stratigraphic sequence and strata thickness of Songshan in western Henan and the Benxi and Jiaodong regions are approximately the same as those in the Huainan region, and they are located on both sides of the same sea basin.

(2) Anhui Compound Syncline

This syncline is also known as the Yangtze River fold-fault zone, as it spreads on both banks of the Yangtze River in the Jiangsu and Anhui regions. It is approximately bounded by the Chuhe fault zone to the northwest. It extends southwest through the Lujiang Kongcheng–Gaobu River to the eastern side of the Qian Mountains and against the Jiangnan Ancient Uplift to the south. The central part of this zone is the Nanjing Yizheng underwater uplift. The early Sinian sediment is less than 400 m thick in the middle but more than 1,000 m thick in the depressions on both sides. The late Sinian sediment is dominated by restricted platform facies composed of dolomites, and this uplift remained at the end of the Cambrian. In the Ordovician–Silurian period, the Chuhe depression on the northern side disappeared and was connected to the riverside zone. Therefore, the depression existed on the south side and accepted shallow sea-coastal facies carbonate and single continental clastic sandy muddy deposits. Since the Late Devonian, this underwater uplift has been complicated by more subuplifts and depressions. It turned into a depression zone in the Late Carboniferous and became the deposition center of the Lower Yangtze Depression.[8] The central part of the Triassic depression became an S-shaped long and narrow trough, which was the only subsidence zone in the Lower Yangtze region.

The fold-fault zone along the Yangtze River formed a fold zone, and both folds and faults developed in the Indochina Movement. Because this zone is located in the composite region of the N-E-trending system and the eastern flank of the Huaiyang epsilon-shaped front arc zone, the most notable feature of the folds is their S-shape, accompanied by the development of compression and compression torsional faults and large-scale Indosinian monzonitic rock masses.

(3) Tectonic Dynamic Metamorphic Zone on the Southeastern Margin of the Dabie–Jiaodong Uplift Fault Zone

This zone spreads along the main fault zones on the southeastern margin of the Dabie–Jiaodong uplift fault zone and features a series of N-E-trending high-pressure and ultrahigh-pressure dynamic metamorphic mineral zones along the fault zone. It starts from the river collapse in Susong County in Anhui Province, spreads northeast through the Qianshan, Yuexi and Tongcheng, and intersects with the Tanlu fault zone near Lujiang County. However, it is still exposed in the corresponding region on the western side of Lujiang County and then passes through the eastern mountain of Feidong, Zhangbaling and Chuzhou and Jiangsu Xuyi, Xinyi, Donghai and intermittently through Shandong Rizhao, Qingdao and Wendeng into the sea, with an intermittent length of more than 1,000 km and a width of several thousand meters to tens of kilometers in the overall direction of 40°–50° north to east, and it is nearly N-N-E-trending in the intersection section of the Tanlu fault zone. Due to the later intersection, transformation and misalignment of the N-N-E-trending system, there are sometimes discontinuities or geese arrays, which are strong fault zones on the southeastern margin of the Dabie–Jiaodong uplift fault zone. The rocks along this fault zone were squeezed and broken into cataclastic rocks, mylonites, schists and lenses, accompanied by blue amphibole, polysilica muscovite, pyroxenes, graphite, chlorite, jadeite, red curtain and other tectonic dynamic metamorphic minerals, i.e., rock, as well as banded blue amphibole schist, polysilica muscovite schist, eclogite and other high-pressure dynamic metamorphic rocks and coesite. Isotopic age test data of these stress minerals have shown that these dynamic metamorphic minerals formed in the Indosinian period, and only a few formed in the Neoproterozoic and Yanshanian periods. Among them,[9] the eclogite Sm-Nd formed at 221–244 Ma. The main formation period of the tectonic dynamic metamorphic rock zone is consistent with the strong tectonic deformation period in the region.

4.1.4 N-E-Trending Tectonic System in South China

In the South China region south of the Qinling–Dabie Mountains, the N-E-trending tectonic system is widely distributed, with large-scale uplift zones as the main body and compound folds and faults as the main tectonic components. There are also granite-mixed rock zones, thermodynamic metamorphic zones and ductile shear zones. The N-E-trending tectonic system can be divided into the Caledonian and Indosinian systems based on the formation period. They are S-shaped, mostly due to the limitation and strong interference of other tectonic systems. In addition, there are some paleo N-E-trending tectonic zones in the pre-Sinian system. The N-E-trending tectonic system in South China includes three first-level uplift zones, i.e., the Longmen Mountain–Yulong Snow Mountain uplift zone, the Tianmu Mountain–Jiuling Mountain–Xuefeng Mountain uplift zone, and the Fujian–Jiangxi–Guangdong uplift zone from west to east, among which the second zone is the largest zone, followed by the third zone. There are large Sichuan–Guizhou Paleozoic fold zones and Hunan–Guangxi–Guangdong–Jiangxi Late Paleozoic fold zones between the three uplift zones, forming a N-E-trending tectonic pattern of "three uplifts and two depressions".

(1) Longmen Mountain–Yulong Snow Mountain Uplift Zone

The zone spreads in an overall N-E-trending direction, passing through Hanzhong to the northeast, intersects with the Qilian–Dabie N-W-W-trending zone, is obliquely mitered by the Qinling E-W-trending zone, passes through Kangding and is reversely compounded by the Danba N-W-trending compound anticline and Sichuan–Yunnan N-S-trending zone to the southwest, passes through the N-S-trending zone of Sichuan-Yunnan, Jinping Mountain, Yulong Snow Mountain and the eastern side of Diancang Mountain to the southwest, and is later reversely compounded with the Ailao Mountain N-W-trending tectonic zone. It traverses the middle part of the Yangtze–Talimu block and is divided into two sections by the Sichuan–Yunnan N-S-trending zone. The northeastern section is called the Longmen Mountain fold-fault zone, and the southwestern section is called the Yanyuan–Lijiang fold-fault zone, which features extensively developed thrust tectonics.

(a) *Longmen Mountain fold-fault zone*: This zone starts from Mian County south of Shaanxi through Guangyuan–Maowen, Guanxian–Yibaoxing and Tianquan and is diagonally inserted into the Sichuan–Yunnan N-S-trending zone to Luding, with a length of 500 km and a width of only 25–40 km. It is composed of N-E-trending folds, torsional faults and compressional fault zones at 40°–50° north to east. The main tectonic traces of this tectonic zone include four compound anticlines, three compound synclines and four large-scale compression and torsion-compression fault zones. The four compound anticlines are the Jiaoziding–Mutuo compound anticline, the Pengguan–Juligang compound anticline, the Baoxing compound anticline and the Tianjingshan compound anticline. The cores of the first three wings are exposed in the pre-Sinian and Jinning–Chengjiang periods as intermediate-acid magmatic rocks, and the two wings are composed of the Sinian–Triassic system in a ξ-shaped oblique array, which is caused by the counterclockwise torsion of the main tectonic zone. The three compound synclines are the Wu'an syncline, the Yanjing–Wulong syncline and the Tangwangzhai–Yangtianwo syncline. The strata involved is the Sinian–Triassic system, and there is no angular unconformability among them. The axis of a single fold is 45°–60° north to east, and there are S-shaped turns at both ends. In addition, the axis of the geese array is at 45° north to east. The four large faults are the Qingchuan fault zone, the Maowen fault zone, the Beichuan–Yingxiu fault zone and the Jiangyou–Guanxian fault zone at 45°–60° north to east, extending approximately 200–400 km. The first three fault zones intersect the Triassic and previous rock formations and rock masses, and the latter zone intersects the Jurassic system, with a section at 50°–80° north to west. The dip angle of the southern section of the Beichuan–Yingxiu fault zone is only 20° and is wavy and curved. The oblique thrust scratches are often covered by horizontal scratches, the faulted tectonic rocks are developed, and the fault zone overlaps the section. Their deformation features and the relationship with the intersected strata suggest that these faults formed mainly in the Indosinian period and were still active in the Yanshan period but showed a thrusting nappe and clockwise horizontal displacement in the Himalayan period (Fig. 4.7). They have a genetic relation with the eastward squeezing of the Qinghai–Tibet Plateau and the activities of the Neoid N-S-trending zones.

(b) *Yanyuan–Lijiang fold-fault zone*: This zone diagonally spreads between the Sichuan–Yunnan N-S-trending zone and the Sanjiang N-S-trending zone. It was originally called the West Yangtze block depression between the Jinhe–Qinghe fault zone and the Xiaojinhe fault zone and has been slowly subsiding time since the Paleozoic, with a well-developed caprock and a maximum thickness of 16,500 m. In particular, the Devonian and Carboniferous systems are well developed, and the basement rock system within the zone is not exposed. Meanwhile, the Sinian-Middle Triassic sediments are mostly the Yangtze type but are different from the Yangtze type since the Late Triassic. From the perspective of

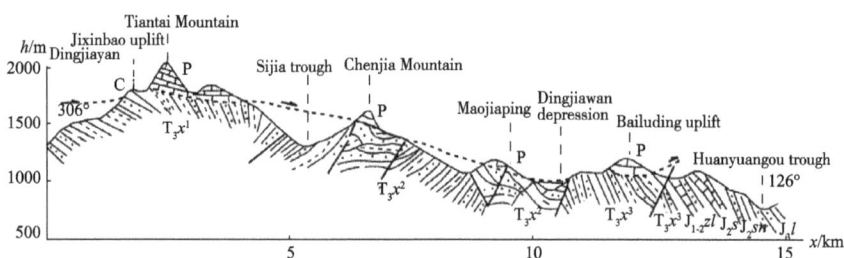

Fig. 4.7.   Tectonic profile of the Tiantai Mountain–Bailuding flying peak in Peng County (Regional Geology of Sichuan Province, 1991).

deformation features, they can be divided into Jinhe–Yongsheng fold faults and Yanyuan–Heqing'ao folds. Based on the analysis of regional formation data, the Yanyuan–Lijiang fold-fault zone was folded and uplifted mainly in the Indosinian Movement and was noticeably affected by the Caledonian Movement, resulting in the absence of a Silurian system at both ends. In addition, the Devonian systems are unconformably parallel to the Sinian Dengying Group in some regions.

(c) *Longmen Mountain–Yulong Snow Mountain fault zone*: This zone generally spreads in the N-E-trending direction, the northeastern end extends into the Hanzhong region and the southwestern section extends to the Ailao Mountain fault zone. Whether its later activities are beyond the Ailao Mountains and extend to southwestern Yunnan is still unclear.

The main components of this fault zone are the Xiaojinhe fault, Beichuan–Yingxiu fault and Maowen fault in the northeastern section and the Xiaojinhe–Sanjiangkou fault, Jinmian–Lijiang fault and Qinghe fault in the southwestern section, all of which are generally deep and large faults in the N-E-trending direction. The Paleozoic–Triassic depositional environment gradually deepened, and the deposition thickness gradually increased from south to north. The Qinghe–Jinhe fault on the southwestern section was strongly active in the Neoproterozoic period, and some strong activities also occurred along several faults in the late Hercynian period. There were basic magma eruptions and intrusions along the Yanqinghe fault, Xiaojinhe–Sanjiangkou fault and Jinmian–Lijiang fault.[10] These features indicate that obvious extensional activities occurred in this group of faults during the Paleozoic-Triassic period. The fault system passes through the Yanyuan region to the northeast and gradually turns northward on the western side of the Sichuan–Yunnan N-S-trending zone in the N-N-E-trending direction. They gradually obliquely intersected and reconnected to the Yalongjiang fault zone of the N-S-trending tectonic system, developing thrust nappe tectonics and flying peaks.

The Longmen Mountain section in northwestern Sichuan includes the Maowen deep fault, Beichuan–Yingxiu deep fault and Jiangyou–Guanxian fault. The Beichuan–Yingxiu deep fault zone was originally called the Longmen Mountain main central fault in the N-E-trending direction, with a length of more than 400 km. The fault zone is multibranched and compounded on the plane, with wavy sections and superimposed nappe rocks sandwiched among them. The basic features of the Longmen Mountain fault zone are as follows: the superimposed rock slices at the rear of the nappe compression zone were subject to ductile shear deformation, the central part was subject to ductile deformation, accompanied by brittle deformation, and the lower rock slices at the front were subject to brittle deformation. From the perspective of deformation features, dynamic metamorphism occurred in the rear part, with a medium-pressure dynamic thermal gradual metamorphic zone and mixed petrification. The separation occurred only along the shear zone in the middle part, with new stress minerals or metamorphic minerals. There is basically no metamorphic effect in the front. From the perspective of kinematics, the axis of the flexible shear folds associated with shear slip, slip shear scratches, gravel stretch lineage, and mineral rotation orientation all consistently indicate thrust-napping activities in the rock fragments from northwest to southeast. From the perspective of overturning time, there was continuous squeezing, migration and stacking process in multiple stages and multiple periods under tectonic stresses from west to east from the end of the Indosinian period to the Yanshanian period and then to the Himalayan period.

The most obvious tectonic feature of the Longmen Mountain–Yulong Snow Mountain tectonic zone is that the nappe tectonics or nappe masses along the tectonic zone are extensively developed. Its main body is basically composed of numerous nappe masses and flying peaks, especially in the Longmen Mountain region, which appears as nappe mountains piled up by nappe masses. Aeromagnetic data indicate that the tectonic zone is located in a regional N-E-trending negative anomaly zone, in which the locally elevated linear anomaly and continuous negative anomaly gradient zone indicate the location of the main fault zone. There is no strong magnetic anomaly in the anomaly zone, and the magnetic anomaly on the exposed magmatic rock body is only 90 nT. It is an isolated and self-loop area, indicating that the magnetic body is limited in depth. The gravity and magnetic profiles across the "Pengguan Compound Rocks" indicate that the compound mass is not deep but is floating and "rootless". The fold deformation and metamorphism of this zone occurred mainly in the Indosinian Movement through squeezing and thrust. The main fault was transformed into a thrust nappe, developed and formed in the Yanshanian period and finalized in the Himalayan period, forming nappe tectonics and flying peaks with varying scales. There are not only Jurassic–Cretaceous molasse formations but also Paleogene–Neogene and Quaternary molasse formations along the front margin of the Longmen Mountain–Yulong Snow Mountain tectonic zone, indicating the long-term and multistage nature of the development of this nappe zone.[11] The thrust zone of the nappe tectonic is mostly W-N-trending plow-like in sections. The inclination angle is generally 50°–60° on the surface and tends to decrease downward, to only 5°–10°. Petroleum seismic data indicate a low-velocity layer at approximately 21 km below the surface of the Longmen Mountain region, which is an ideal main slip surface. Based on the current positions of the flying peaks and their root zone positions with similar lithology and age, it is estimated that the maximum displacement distance was 30–40 km. Because the zone nappe occurred in the shallow part of the crust and there was no deep tectonic background there, there was no corresponding magmatic activity or hypermetamorphism. However, there was mid-high-pressure dynamic thermal metamorphism, mixed petrification and stress minerals in the rear and middle of the main fault zone. In the Qingchuan Great Fault, there is a high-pressure metamorphic rock zone composed of blue amphibole facies and a schistally distributed granite vein group zone with an isotopic age of 223 Ma, which are also the products of tectonic dynamic metamorphism.

(2) Tianmu Mountain–Jiuling Mountain–Xuefeng Mountain Uplift Zone

This zone is the main body of the South China tectonic system, including the Jiangnan uplift zone and the fold-fault zones on its sides. The N-E-trending tectonic zone dominated by Paleozoic folds (including some Neoproterozoic folds) is well preserved as a more continuous reverse S-shaped N-E-trending zone. The core of the uplift zone is composed of pre-Sinian shallow metamorphic rocks, and its two wings are composed of Paleozoic and Early Middle Triassic strata, which are embedded under the South Jiangsu Plain on the east and continuously extend under the Yellow Sea and the East China Sea. The buried Suzhou N-E-trending fault zone is an important component of the uplift zone. The zone spreads southwest through the central Guizhou region. The N-E-trending tectonics in this uplift zone consist of late N-E-trending tectonics finalized in the Indosinian period and early N-E-trending tectonics in the Lower Paleozoic. Both have the same axis and similar deformation, but they have different intensities with the boundary of the Caledonian Movement. In addition, the zone has also merged with the Sino-New Proterozoic Paleo-N-E-trending tectonic zone, with similar orientations and obvious inheritance activities. In the Early Paleozoic, the uplift zone was the boundary between the active deposition region (trough region) on the southeastern coast and the stable deposition region on the west (platform region). At the beginning of the Late Paleozoic, it was also the northwestern boundary of the Cathaysia Paleo-Continent on the southeast. Due to severe interference from other tectonic systems in the later period, it features deflected orientation, a unique shape, uncoordinated wings and even covered and disconnected area, forming a NE-NEE-N-E-trending reverse S-shaped tectonic system. The main components are as follows.

(a) *Tianmu Mountain–Huaiyu Mountain uplift zone*: This zone is the northeastern section of the Tianmu Mountain–Jiuling Mountain–Xuefeng Mountain uplift fold zone and is located in the Huaiyu Mountain and Tianmu Mountain regions adjacent to Jiangsu, Anhui and Jiangxi, which are composed of the Tianmu Mountain–Huaiyu Mountain compound uplift zone, a series of fault zones and some granite rock masses, including the Anji–Qimen–Jingdezhen compound anticline zone, the Hangzhou–Kaihua compound syncline zone on the southern and northern sides, and three first-level tectonic zones in the Xuancheng–Dongzhi syncline zone. The Anji–Qimen–Jingdezhen compound anticline zone is N-E-trending, and its axis is approximately located in the Qimen–Jingdezhen region. It is composed of the Mesoproterozoic Banxi Group, Shuangqiao Mountain Group shallow metamorphic rocks and Sinian system, and its two wings are composed of the lower Paleozoic and Middle-Lower Triassic systems. The Paleozoic Tianmu Mountain compound anticline is its eastern extension and is then submerged under the Jiangsu Plain on the east. The Xuancheng–Dongzhi compound syncline zone is composed of a series of open folds, such as the Pailou compound syncline, Qidu compound anticline and Taiping compound syncline, with the axis at 40°–60° north to east. It is composed mainly of Paleozoic systems. The anticline axis is the Precambrian or Cambrian system, and the syncline core is the Silurian, Permian or Triassic system. The Hangzhou–Kaihua compound syncline zone is composed of the Hangzhou–Kaihua compound syncline at 50° north to east, the Linpu–Shouchang and Lanxi–Jiangshan compound synclines to the south, and the Wukang–Lu village and Changxing–Xiaofeng compound synclines to the north. The N-E-trending tectonic traces with similar axial directions in the early and late phases are composed of the Lower Paleozoic, the Upper Paleozoic, and the Middle-Lower Triassic systems.

The regional faults associated with the uplift zone are also well developed, including mainly deep and large faults, i.e., the Jiangnan,

Anhui–Zhejiang–Juangxi, Northeast Jiangxi and Jiangshan–Shaoxing faults. The faults are mainly E-S-trending, with obvious activities and features of thrust nappe and slip nappe in multiple periods. The Jiangnan and Jiangshan–Shaoxing fault zones are taken as examples.

The Jiangnan large fault zone passes diagonally through the mountainous region of southern Anhui, Xuancheng and Dongzhi, connects with the Xiushui–De'an deep fault in Jiangxi on the west, extends into the Suyang region of Jiangsu on the north, and intersects with the Indosinian intrusive rocks. The aeromagnetic anomaly progressive zone spreads along the fault zone. It was formed in the Caledonian period and had obvious control effects on the Paleozoic lithofacies, thickness and biota, and it was finalized in the Indosinian period.

The Jiangshan–Shaoxing deep fault zone is large, with a length of 280 km and a width of 10 km. It has a slightly S-shaped anomality zone with a width of several kilometers. There is deformation and metamorphism and strong magmatic activity along the fault zone from multiple periods, but Paleozoic–Early Mesozoic activity occurred in the N-E-trending tectonic system.[12] There are strong ductile deformation and developed thrust faults in the Shuangqiao Mountains on the northwestern side of the fault zone and the Chencai Group on the southeastern side of the fault zone, forming imbricated thrust tectonic zones. The zones include mixed quartz diorite zones with a width of 10 km and schistological mixed long granite zones and Qianmi rock zones, as well as a large number of lenticular ultrabasic rocks. The fault is regarded mostly as the product of the collision between the Jiangnan and Cathaysia paleo-continental blocks in the Jinning and Caledonian periods. Notably, the abovementioned N-E-trending compressional tectonics in the pre-Sinian metamorphic basement covered unconformably by the Zhitang Formation are different from the Paleo-N-E-trending tectonic zones formed in the early and late Cathaysian periods.

(b) *Mufu Mountain–Juling Mountain uplift zone*: This zone is the middle section of the Tianmu Mountain–Jiuling Mountain–Xuefeng Mountain uplift zone. The main body is the uplift zone composed of the Mufu Mountain compound anticline zone and the Jiuling Mountain compound anticline zone, and its southern and northern sections are the Pingxiang–Leping compound syncline zone and the Hukou–Tongshan compound syncline zone, respectively. The axis of the compound anticline is composed of the Meso-New Proterozoic shallow metamorphic rock system and granitic complex rocks, and its two wings are composed of Paleozoic and Lower Triassic systems. The axis of the compound syncline is composed mainly of the Lower Triassic system, and its two wings are composed of the Paleozoic system. The Mufu Mountain compound anticline zone is NE-NEE-trending, with slight convexity to the northwest and the Wugong Mountain compound anticline to the south.

On the side of the Mufu Mountain–Jiuling Mountain uplift zone, the same trending fault zones are highly developed and represented by the Xiushui–De'an, Yifeng–Nanchang, Pingxiang–Guangfeng and other fault zones. They are obviously active in multiple periods, and some fault zones show strong thrust and slip features. The Xiushui–De'an fault zone is located on the northern margin of the Jiuling uplift zone, exhibiting developed tectonic complex blocks with obvious ductile deformation, which controls the distribution of the Paleozoic deposits and the Yanshanian complex rock zones. The Yifeng–Nanchang fault zone is located on the southern margin of the Jiuling uplift zone, at greater than 60° north to west, and it is composed of a series of stacked thrust faults, with obvious brittle and ductile deformation. It controls the distribution of the Mesoproterozoic volcanic rock zones and Mesozoic and Cenozoic basic and ultrabasic compound zones. The Yifeng–Nanchang fault zone and the Shanggao Qibaoshan–Gaoanxinjie thrust fault zone on the south moved southward in the Indosinian–Yanshan period, forming a double-zone thrust fan or double thrust tectonics, i.e., the thrust nappe tectonic on the southern margin of Jiuling Mountain (Zhu Zhicheng, 1987). The Pingxiang–Guangfeng fault zone is located on the northern margin of the Wugong Mountain uplift zone, and it forms a boundary between the Yangtze block and the South China fold system, which corresponds to the deep tectonic variation zone. It has a significant control effect on the deposition thickness of the Neoproterozoic and Lower Paleozoic systems. In addition, there are basic and ultrabasic rock masses.

(c) *Xuefeng Mountain–Leigong Mountain uplift zone*: This zone is located in the southwestern section of the Tianmu Mountain–Jiuling Mountain–Xuefeng Mountain uplift zone and is dominated by large uplift zones. The same-trending fault zones are also well developed and separated by the Yuanma Basin. The northeastern section is combined with the E-W-trending regional tectonic zone, forming the large "Xuefeng arc-shaped tectonic zone" protruding northwest. It can also be divided into early and late phases. In the early period, the N-E-trending tectonic system was composed of the Meso-New Proterozoic and Lower Paleozoic system with clear traces. In the late period, the N-E-trending tectonic system developed in the depression zone of the early N-E-trending tectonic system. It is composed of the Upper Paleozoic and the middle-lower Triassic systems. Due to the interference and destruction of the N-N-E-trending tectonic system, its traces are mostly incomplete. From northeast to southwest, there are four secondary tectonic zones, including the Wuling Mountain fold-fault zone, Anhua–Taojiang fold-fault zone, Southeast Guizhou fold-fault zone and Xuefeng Mountain fold-fault zone.[13]

Qiu Yuanxi *et al.* studied the strato-slip effects in the Jiangnan–Xuefeng region and discovered and determined the Caledonian fold nappe tectonics in the Hunan–Guangxi–Guizhou region adjacent to the Xuefeng uplift zone. The tectonic pattern is presented by mainly near-horizontal folds and overburdened imbricated tectonic fans, dominated by the Banxi Group and including the upper wing of a giant horizontal fold zone composed of Sinian, Cambrian and Ordovician systems, with a width of approximately 300 km and covered unconformably by the Upper Paleozoic system. Its frontal zone is located between Dushan–Shibing and Kaili Guiding–Zhenyuan, with a series of S-E-trending imbricated thrust zones. For example, from Zhenyuan Shidongkou to Xincheng Liangtuao, there are 5–6 S-E-dipping N-E-trending reverse fault zones, forming an imbricated tectonic fan westward. Among them, the Banxi Group exposed in the Shidongkou faults is overridden on the Cambrian system, with flying peaks and tectonic windows. Qiu Yuanxi *et al.* believed that the continental margin fold mountain system and fold thrust (slip) nappe tectonics formed by the collision and squeezing of the southeastern margin of the Caledonian Yangtze block. Therefore, the layer slip effects in this period are mainly the product of the early N-E-trending tectonic stress field.

(3) Fujian-Jiangxi–Guangdong Uplift Fold Zone (Wuyi–Yunkai Uplift Fold Zone)

This zone spreads in western Fujian, eastern Jiangxi, western Guangdong, southeastern Guangxi and other regions. It is composed mainly of compound anticline zones with pre-Sinian and Lower Paleozoic metamorphic rocks and mixed rocks as the cores and compound syncline and fault zones with the Upper Paleozoic system as the cores. Due to the compound transformation of the Nanling E-W-trending tectonic zones and the strong interference of the N-E-trending and N-N-E-trending tectonics, it is discontinuously reverse S-shaped, forming a giant uplift zone parallel to the Huaiyu–Jiuling–Xuefeng large uplift zone. From West Yong'an to Meixian County in eastern Guangdong, there is a Hercynian–Indosinian depression zone superimposed on the Caledonian fold zone, which is commonly called the Yongmei Depression. The main body is N-E-trending, and a layer of single continental debris formation combined with coal-bearing formation and carbonate rock formation with a thickness of 4,500 m was deposited in the late Paleozoic. A transitional fold was composed of the Upper Paleozoic system, and there was a Yongmei thermodynamic metamorphic zone.

The Yunkai uplift fold zone is the southwestern section of the Fujian–Jiangxi–Yuelong fold zone. It is located in the Yunkai Mountain–Liuwan Mountain region in western Guangdong and southeastern Guangxi. The main body is a compound anticline zone composed of Sinian and Early Paleozoic metamorphic rocks and mixed rocks, and there are also N-E-trending Paleozoic folds and some deep and large fault zones on both sides, including the Dayunwu Mountain compound anticline zone on the east, the Liuwan Mountain compound anticline zone on the west, and the Lower Paleozoic compound syncline zone on the side. The Dayunwu Mountain compound anticline zone (known as the "Yunkai Uplift") and its side deep faults have obvious control over the Paleozoic deposits. There are mainly active flysch deposits with a thickness of approximately 10,000 m in the Pre-Sinian and Lower Paleozoic periods. In the late Caledonian period, the zone was gradually uplifted, and intense regional metamorphism, migmatization and magmatism occurred. The central uplift zone (the Liuwan Mountain uplift) of the Qinzhou Hercyn Trough is composed of the Liuwan Mountain compound anticline zone, which controls the distribution of the Liuwan Mountain Indosinian granite zone. Flysch-like formations with a thickness of 10,000 m were deposited in the Qinzhou–Lingshan region and Bobai–Cenxi region on both sides in the Silurian-early Permian period, forming two N-E-trending depression zones. There are also some N-E-trending syncline zones on the side of the Yunkai Uplift, mainly including Yangchun Bay Gaoyao-Qingyuan, Huaxian and other synclines, and they are composed of the Upper Paleozoic system and are components of the Yunkai N-E-trending tectonic system.

In addition, the N-E-trending deep faults are well developed on the sides of the Yunkai Uplift and the Shiwan Mountain Uplift, i.e., the Wuchuan–Sihui, Cenxi–Bobai, Lingshan–Qinzhou and other deep faults. They were obviously active in multiple periods, with ductile deformation, thermodynamic metamorphism, and multiphase slips and detachments. They not only have an obvious control effect on the Paleozoic, Caledonian and Indosinian granites but also control the genetic development of the Jurassic and Cretaceous basins. However, their activities are presented by N-E-trending tectonics in the early and late Paleozoic, which have merged into the eastern wing of the Huanan epsilon-shaped front arc zone since the Mesozoic.

The Qiongzhong uplift fold zone through the central part of Hainan Island may be an extension of the Yunkai uplift fold zone. The main body is the N-E-trending Qiongzhong compound anticline and is composed mainly of Proterozoic and early Paleozoic metamorphic rocks. The axis is located in the Anding, Qiongzhong and Ledong regions. Because it is invaded by the Caledonian–Hercyian Qiongzhong mixed granites and the Yanshanian granites, the preserved folds are extremely incomplete and represented by the N-E-trending South Kunyuan syncline and the Danxian syncline.

In the two first-level fold zones among the three first-level uplift zones on the South China N-E-trending tectonic system, the N-E-trending tectonics are also well developed. The N-E-trending tectonic system in the Sichuan-Guizhou Paleozoic fold zone is mostly compounded and used by the N-E trending systems. However, the Huaying Mountain and Qiyao Mountain N-E-trending fault zones are the main bodies of the N-E-trending system of the depression zone and have played an important role in the establishment of the N-E-trending Paleozoic Sichuan Basin.[14] The Early and Late Hercynian N-E-trending tectonics can be seen in the Hunan–Guangxi–Guangdong–Jiangxi Late Paleozoic fold zone in the southern Hunan–northeastern Guangxi and northwestern Guangxi regions. For example, the closed N-E-trending fold zone composed of Cambrian shallow metamorphic rocks can be seen in the Xiaoshui Basin, which is made unconformable by the Devonian system. The wide and gentle Upper Paleozoic folds are also relatively developed and accompanied by faults. The above Caledonian movement planes are two types of N-E-trending folds and associated faults based on the dividing plane, which should be the components of the early and late N-E-trending tectonics, respectively.

In summary, the N-E-trending tectonic system in East China has the following basic features:

(1) The N-E-trending tectonic system features plastic deformation, and N-E-trending large-scale compound uplift zones and depression zones are often present, which are accompanied by large compression and compression torsion fault zones with inherited features, as well as intermediate-acid intrusive rock zones and dynamic metamorphic zones.

(2) The N-E-trending tectonic system was finalized mainly in the Indosinian Movement and can be divided into early and late phases in the Jiangnan–Xuefeng uplift zone and its southeastern region in South China. In the early phase, they were the N-E-trending uplift zones and depression zones that formed in the late Caledonian Movement, accompanied by granite zones and dynamic metamorphic zones, distributed mainly in the Wuyishan–Yunkai uplift fold zone. In the late phase, they formed in the Indosinian Movement, and there were angular unconformably sequences between the upper and Lower Paleozoic systems. The directions of the tectonic lines of the two fold zones are basically the same. In the Yangtze region, North China and the broad region of Northeast China, uplifts dominated the Caledonian Movement, without obvious deformation, metamorphism and magmatic intrusion. A wide range of N-E-trending uplift and fold zones formed in East China in the Indosinian Movement, with approximately the same deformation and metamorphism. Due to the superimposed effect of deformation and metamorphism on the western phase in South China, the N-E-trending system is strong on the east and south and weak on the west and weak north.

(3) The tectonic dynamic metamorphic zones and intrusive rock zones are distributed unevenly. Due to the effects of N-E-trending tectonic waves and different blocks, strong tectonic dynamic metamorphic zones and intermediate-acid intrusive rock zones often formed between different blocks in the affected region, creating dynamic metamorphic zones, i.e., mixed rock zones and granite zones on the western side of the Wuyishan–Yunkai uplift fold zone, the East Dabie–Jiaonan high-pressure and ultrahigh-pressure dynamic metamorphic zone, Zhangguangcailing–Laoyeling tectonic magma zone, etc.

At the same time, large fault zones with ductile shear features and inherited activities are more common and have a significant control on the formation of the Paleozoic and Early-middle Triassic lithofacies.

(4) The N-E-trending tectonic system has been subjected mostly to strong compound transformations. Its formation was affected by the compound transformation of the late tectonic systems, resulting in its deformation and misalignment. Therefore, its position was peculiar or intermittent or formed an S-shaped, inverted S-shaped or arc-shaped distribution, i.e., Xuefeng arc, Jiuling arc, Yangchun S-shaped compound syncline, etc. In addition, the formation of the N-E-trending uplift and fold zones have been subject to many strong uplifts, resulting in multistage and multilevel detachments on the side, which often show the slip of the uplift zone to the side depression zone. However, these zones are dominated by push-slip action directed to the northwestern foreland depression, forming the Xuefeng, Jiuling Longmen Mountains and other arc nappe tectonic zones.

(5) The N-E-trending tectonic zone has an important control effect on the Paleozoic and Early middle Triassic lithofacies and deposition minerals, the Caledonian and Indosinian magmatic zones and related endogenous and metamorphic minerals, i.e., the Paleozoic iron, coal, marine oil and gas and other mineral fields. There are oil and gas indications in the South China Paleozoic and Middle–Lower Triassic neritic deposits controlled by the N-E-trending tectonic troughs, among which the gentle uplift zones with little tectonic changes and the cap zones with Mesozoic overthrust tectonics may be favorable marine oil and gas accumulation zones.

(6) The main tectonic zones (zones) of the N-E-trending tectonic system are reflected clearly in the deep tectonics, but they are mainly the N-E-trending mantle uplift regions corresponding to the mantle slope zones. In particular, the arc-shaped mantle uplift zones corresponding to the Xuefeng–Yijiuling N-E-trending arc-shaped uplift zone have crustal thicknesses of 30–40 km, indicating that they are regional tectonic systems penetrating the Earth's crust.[15]

This tectonic system should also exist to varying degrees in other blocks, but its determination is beyond the scope of this research.

4.2 New Zealand N-E-Trending Tectonic System in the Pacific Ocean

The system is located at the intersection of the Pacific Ocean block and the Australian block and is composed of the Tonga Trench in the

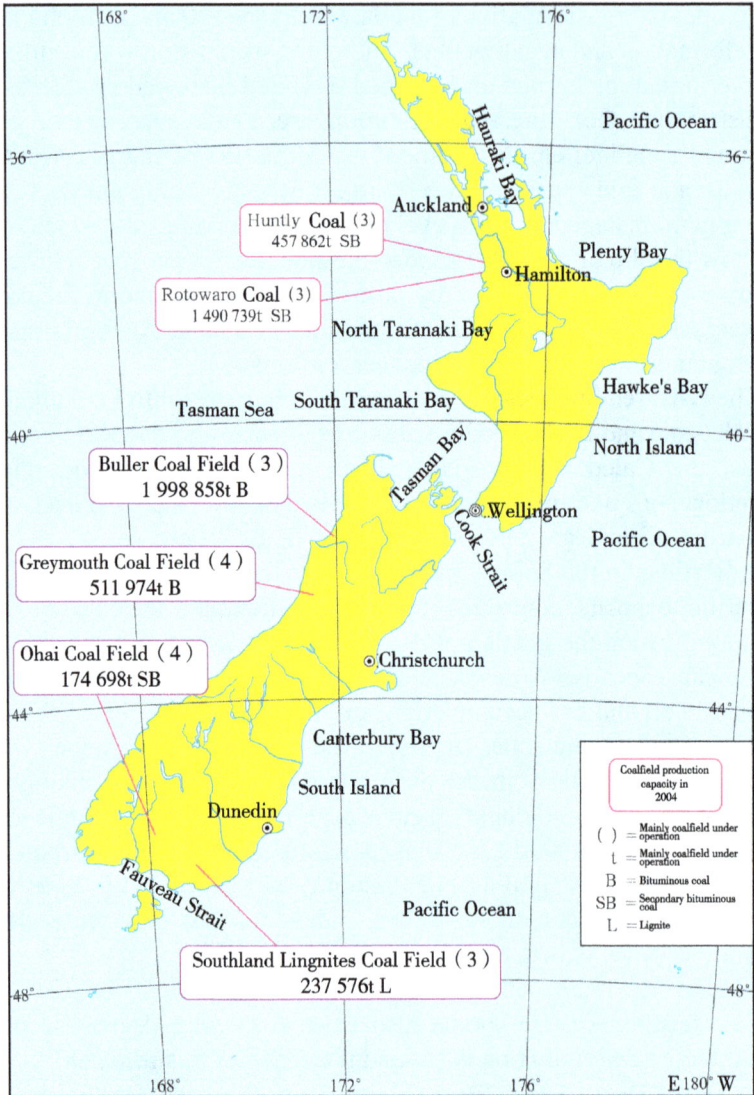

Fig. 4.8.   Distribution of the main coal fields in New Zealand.

Table 4.1.   Main coal field types and output statistics in New Zealand (2004). (Unit: t)

Region	Bituminous coal	Subbituminous coal	Lignite coal	Open-pit coal	Underground coal	Total
Waikato	0	2,053,707	0	1,611,739	441,968	2,053,707
North Region	0	2,553,707	0	1,611,739	441,968	2,053,707
West Coast	2,526,613	100,175	0	2,359,212	267,576	2,626,788
Canterbury	0	3,722	0	3,722	0	3,722
Otago	0	57,051	1853	58,904	0	58,904
South Region	0	174,698	237576	394,988	17,286	412,274
South Island	2,526,613	335,646	239429	239,429	284,862	3,101,688
New Zealand	2,526,613	2,389,353	239429	239,429	726,830	5,155,395

New Zealand subduction zone. The island of New Zealand is composed of North Island and South Island, in which N-E-trending compression and compression torsion faults have developed and include volcanic rock zones (Fig. 4.8). There are very rich coal mines in New Zealand (Table 4.1).

4.3 N-E-Trending Tectonic System in Northwestern Europe

This tectonic system is located on the northwestern margin of Europe, and a series of N-E-trending Mesozoic and Cenozoic basins developed as follows: the Voling Basin, Hebrides Sea, Midland Valley, etc.,[16] accompanied by fault zones and uplift zones (Fig. 4.9).

4.4 N-E-Trending Tectonic System in the Eastern United States

This tectonic system is located in the eastern United States and is composed of multiple Mesozoic and Cenozoic basins, i.e., St. Lawrence, Sizihe, East Coast, Quebec Abracea and other basins and tectonic uplift zones, accompanied by a series of N-E-trending fault zones.[17] The main tectonic zones and fault zones in all basins are N-E-trending (Fig. 4.10).

Fig. 4.9. Distribution of N-E-trending tectonic systems and basins in northwestern Europe.

4.5 N-E-Trending Tectonic System in Eastern South America

This tectonic system is located in the offshore region of eastern South America and is composed of a series of Mesozoic and Cenozoic basins, i.e., the Pudiguar, Shichang Reno, Caokang Changwo, Abrojos,

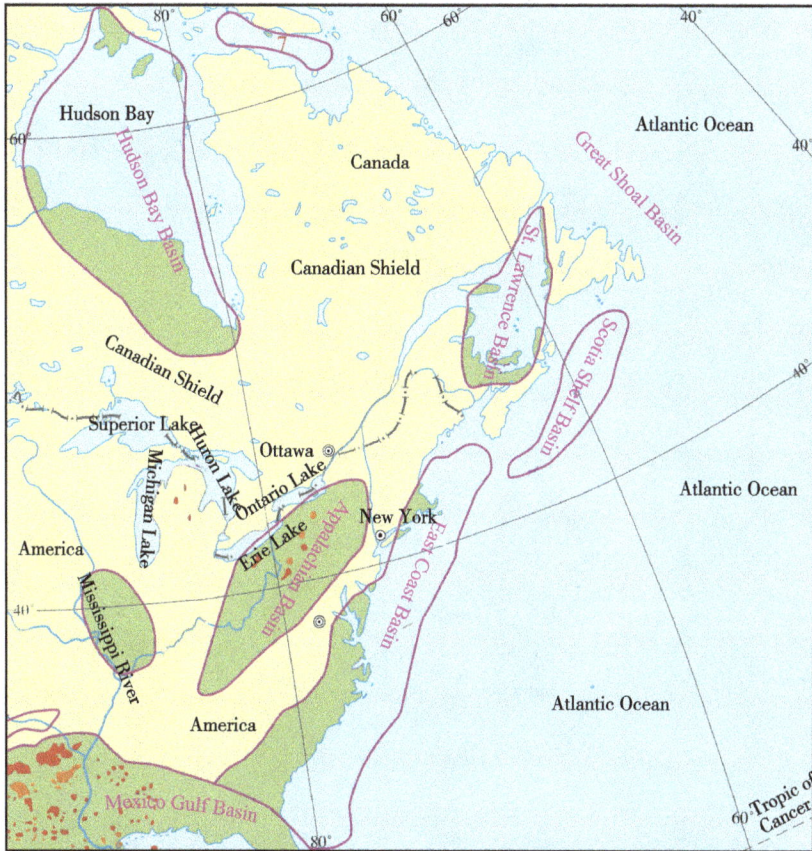

Fig. 4.10. Distribution of N-E-trending tectonic systems and basins in the eastern United States.

Camps Santos and Pilotas basins.[18] There are N-E-trending fault zones, tectonic zones and magmatic rock active zones in the main basins, forming a complex N-E-trending tectonic system.

4.6 N-E-Trending Tectonic System in East Africa

This tectonic system is located in southeastern Africa and is composed of several Mesozoic and Cenozoic N-E-trending basins and the Gadagas tectonic uplift zone.[19] There are N-E-trending fault zones and tectonic zones on the margins of all basins and in the basins. The main basins in the system are include the East Africa Basin, Kazambique Basin, Majunga Basin, Mulundawa Basin, East Udagaska Basin, etc. (Fig. 4.11).

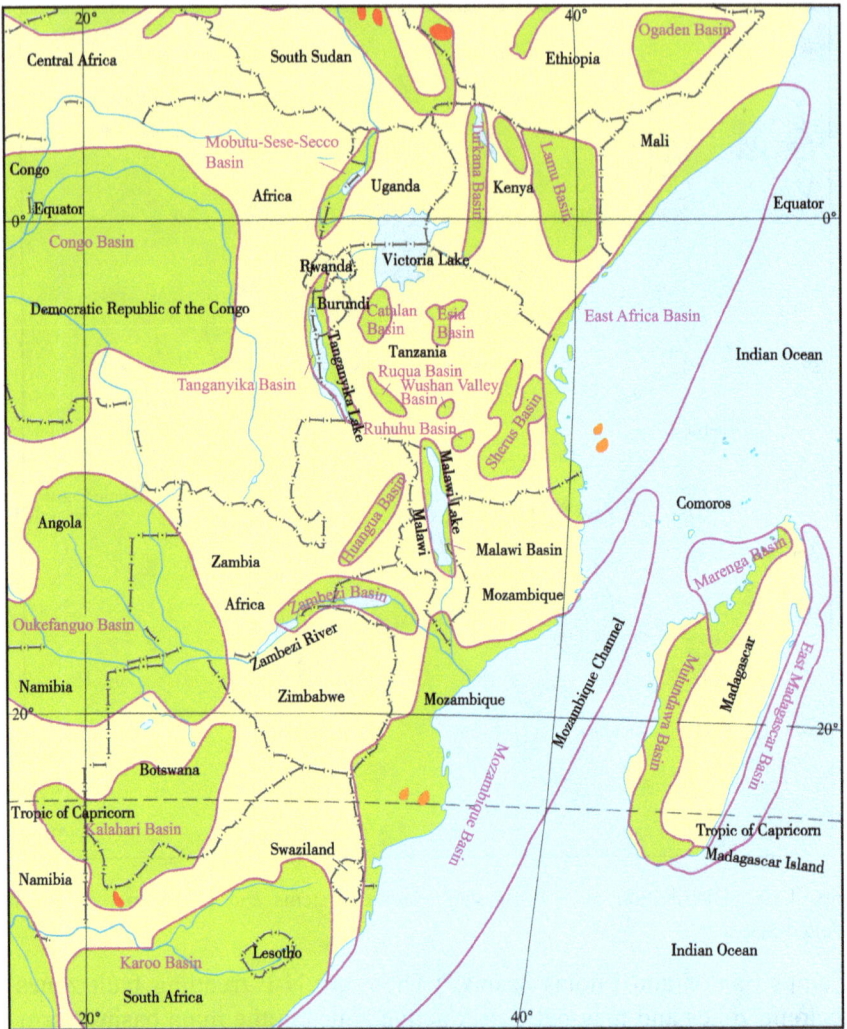

Fig. 4.11.   Distribution of N-E-trending tectonic systems and basins in East Africa.

References

1. Li Dongxu *et al. Introduction to Geomechanics.* Beijing: Geological Publishing House, 1986.
2. Zhang Guojun, Kuang Jun. Petroleum Geological Characteristics and Oil Exploration Prospects in the Hinterland of Junggar Basin. *Xinjiang Petroleum Geology*, 1993, 14(3): 201–208.

3. Zhou Zhiwu *et al.* Geological Structural Characteristics and Petroleum-bearing Properties of the East China Sea. Zhu Xia, Xu Wang (eds.), *Mesozoic and Cenozoic Sedimentary Basins in China.* Beijing: Petroleum Industry Press, 1990.
4. Jin Qinghuan. *Geology and Oil and Gas Resources in the South China Sea.* Beijing: Geological Publishing House, 1988.
5. Zhao Bai. Structural Characteristics and Division of Junggar Basin. *Xinjiang Petroleum Geology,* 1993, (3): 209–216.
6. Chao Jianyi *et al.* Geological Basis and Oil and Gas Enrichment in Bohai Bay Basin. Zhu Xia, Xu Wang (eds.), *Mesozoic and Cenozoic Sedimentary Basins in China.* Beijing: Petroleum Industry Press, 1990.
7. Yuzhu Kang. Petroleum Geological Characteristics and Oil and Gas Prospects in the Northwest of China. *Petroleum Geology & Experiment,* 1984, 6(3): 229–240.
8. Yuzhu Kang. Discovery of Shasheng 2# High-yield Oil Gas Flow and Future Oil Prospecting Direction. *Oil & Gas Geology,* 1985, 6(Supplement).
9. Yuzhu Kang, Zhijiang Kang. Significant Progress in Geomechanics in the Oil and Gas Exploration of the Tarim Basin. *Journal of Geomechanics,* 1995, 1(2): 1.
10. Guo Zhengwu *et al. Research on the Formation and Evolution of the Sichuan Basin.* Beijing: Geological Publishing House, 1996.
11. Geological Mechanics Research Mapping Group, Academy of Sciences. *Specification of Tectonic System Map of the People's Republic of China (1:4,000,000).* Beijing: Geological Publishing House, 1978.
12. Gansu Compilation Group of Stratigraphic Tables. *Regional Stratigraphic Table of the Northwest China · Gansu Volume.* Beijing: Geological Publishing House, 1980.
13. Guan Shicong *et al. Marine Sedimentary Facies and Oil and Gas in China's Seas and Land Transition Areas.* Beijing: Science Press, 1984.
14. Guo Zhengwu *et al. Research on the Formation and Evolution of the Sichuan Basin.* Beijing: Geological Publishing House, 1996.
15. Huang Jiqing. *Research on the Tectonic Features of China.* Beijing: Geological Publishing House, 1984.
16. Jin Qinghuan. *Geology and Oil and Gas Resources in the South China Sea.* Beijing: Geological Publishing House, 1988.
17. Asthana, M., Dubey, S. C. Identification of Thin Sand Bodies within Barail Coal Shale Unit of Upper Assam-A Step Toward Subtle Trap Exploration. *Bulletin of Oil and Natural Gas Omission (India),* 1986, 23(2): 147–159.
18. Austin, J. A., Uchupi, E. Continental-Oceanic Crustal Transition off Southwest Africa. *AAPG Bulletin,* 1982, 66: 1328–1347.
19. Avbovbo, A. A. Tertiary Lithostratigraphy of the Niger Delta. *AAPG Bulletin,* 1978, 62: 295–306.
20. Zhu Zhicheng. Extensional Structures and Detachment Faults. *Geological Technology Information,* 1987, 6(1): 18–24.

© 2024 World Scientific Publishing Company
https://doi.org/10.1142/9789811285561_0005

Chapter 5

N-N-E-Trending Tectonic Systems

**Yuzhu Kang, Shuwen Xing, Zhihong Kang, Yue Zhao,
Zhihu Ling, Zhijiang Kang, and Huijun Li**

Abstract

In this chapter, N-N-E-trending tectonic systems are introduced, including those in China, New Zealand–Tonga, the Eastern United States and the east coast of South America.

Keywords: Tectonic system; N-N-E-trending; Type

This tectonic system (formerly known as the New Cathaysia Tectonic System in China) is composed of the typical N-N-E-trending tectonic zones and deposition zones in East China and is developed in the North American, South American and Pacific regions.

5.1 N-N-E-Trending Tectonic System in China

This tectonic system is a magnificent ξ-shaped tectonic zone in East China and East Asia near the Pacific region. In general, this tectonic system is composed of 3 N-N-E-trending (18°–25° north to east) large uplift zones and large subsidence zones. The topography in East China is obviously controlled by the N-N-E-trending tectonic zones.

The first first-level uplift zone to the east of the N-N-E-trending tectonic system is a strong fold zone on the margin of the East Asian continent near the Pacific Ocean. It is composed of the dry islands in Russia, the Japanese Islands and the Ryukyu Islands in Japan, Chinese Taiwan, Luzon Island and the Balawang Islands in the Philippines from north to south, and the mountains through Malaysia and Kalimantan Island in Indonesia from northeast to southwest. There is a deep sea trench fold zone with approximately the same trend as the uplift zone on the eastern side and a large subsidence zone composed of the Okhotsk Sea, the Japan Sea, the Yellow Sea, the East China Sea and the South China Sea on the western side of this uplift zone. The second first-level uplift zone is composed of the Xihotaling Mountain zone, the tight fold zone that trends diagonally across the Korean Peninsula, and Wuyi Mountain in the southeastern region of China from north to south. Immediately west of this uplift zone is the second-largest subsidence zone composed of large tectonic basins, i.e., the Songliao Plain (including the lower reaches of Heilongjiang), the North China Plain (including the lower Liaohe River and Bohai regions), the Jianghan Plain and the Beibu Gulf south of the Nanling E-W-trending zone to the west of this uplift zone. The third first-level uplift zone is composed of Daxinganling in the northernmost section, Taihang Mountains in the middle section and the fold zone in the eastern part of the Guizhou Plateau in the southern section. The third large subsidence zone is composed of the Hulunbuir–Bayin Heshuo Basin, the Shaanxi–Gansu–Ningxia Basin and the Sichuan Basin to the west of this large uplift zone.[1] There may be a fourth uplift zone of the N-N-E-trending tectonic system to the west of this subsidence zone. It starts roughly from the Olenioclone in Russia and passes southward through the outer Baikal fold zone and the Wendulhan region in Mongolia and the Helan Mountains and into the Longmen Mountains in western Sichuan. The northern section of this uplift zone is arranged in a ξ-shaped pattern, and the southern section is located in the Helan Mountain and Longmen Mountain regions. Due to the strong interference and obstruction of other large tectonic systems (especially reverse S-shaped tectonic systems), it has a weak presentation, and there is a gradual transition to other tectonic systems (Fig. 5.1).

5.1.1 Tectonic System Features

This large tectonic system includes not only the pre-Paleozoic, Paleozoic, some Mesozoic strata and some large intrusive rock masses that occurred

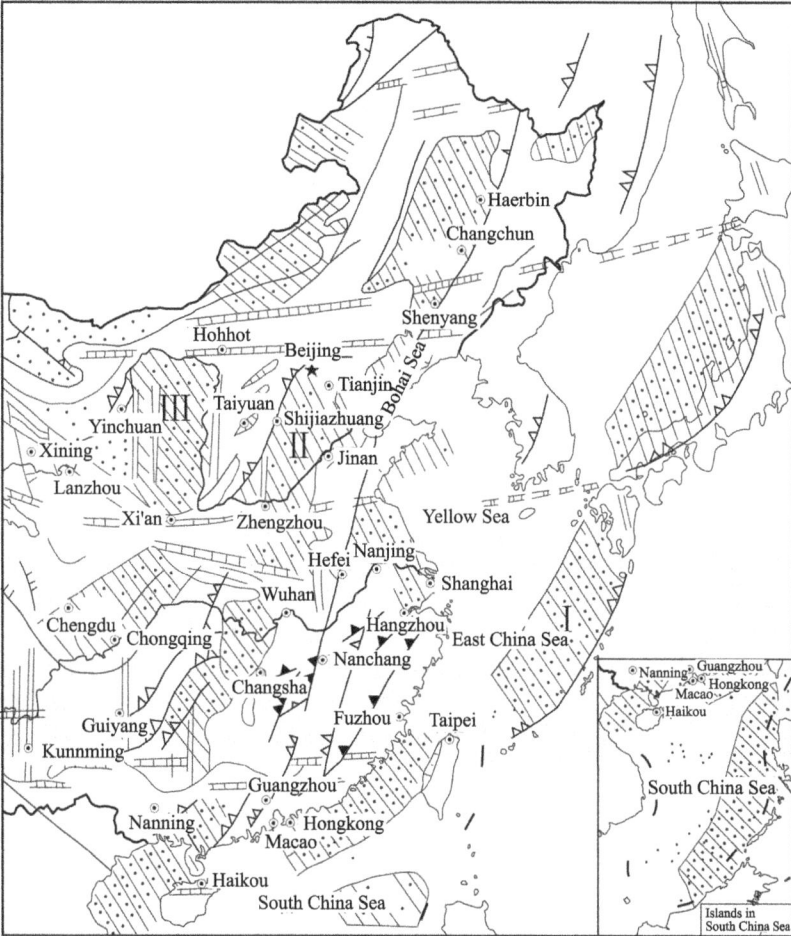

Fig. 5.1. N-N-E-trending tectonic systems in China.

Note: (I) First subsidence zone; (II) Second subsidence zone; (III) Third subsidence zone.

in the early stage in the large regions of eastern China but also large-scale intrusive rock masses and multiphase volcanic rock systems. Therefore, in terms of composition, not only are these large fold zones different but the various sections of each zone are also obviously different.

The first uplift zone on the east, the-Pacific Island arc zone, is extremely complicated, and nearly every section has its own evolution history. Taiwan Province, China, is located in the middle of this uplift zone. Except for the Paleozoic strata with shallow metamorphism exposed to the east, the Paleogene–Neogene strata are extremely widespread.

The western wing of the northern section of the second uplift zone is located mostly in China, and extends from the lower reaches of Heilongjiang to the eastern mountains of Heilongjiang, Jilin and Liaoning, and then to the mountainous region in the Shandong Peninsula or eastern Shandong. It is dominated by large pre-Sinian strata and granite masses, followed by extensively exposed late Paleozoic large granite rock foundations and Yanshanian granite masses. The southern section of the second uplift zone is also more complicated. The axis is composed of a pre-Sinian medium-shallow metamorphic rock system in the Banxi Group and Jianfan Group. The western section is basically a Paleozoic depositional rock system, while the eastern section is widely distributed with Mesozoic, Late Jurassic, Cretaceous acid and intermediate acid volcanic rocks.[2] The intrusive granites in the Yanshanian period are intertwined vertically and horizontally in this uplift zone.

The northern section of the third uplift zone of the N-N-E-trending tectonic system is located mainly in the Daxinganling Mountains and the features in this region are as follows. The late Paleozoic Hercynian granites and the Mesozoic Late Jurassic and Cretaceous intermediate-acid volcanic rocks are widely distributed. There are also some Paleozoic strata in small fragments and some Yanshanian granites. In the middle section is a pre-Sinian deep-middle metamorphic rock system and the overlying relatively gentle Paleozoic rock strata in the vast region of the Taihang Mountains, and there are early Mesozoic Triassic red rock strata in some sections. However, there are nearly no igneous rock intrusions. The southern section is similar to the middle section of the third uplift zone. In the regions of western Hunan, eastern Guizhou, and northern Guangxi, apart from the widespread distribution of the Pre-Sinian Banxi Group, a set of Paleozoic depositional rock systems is also commonly exposed, but there is nearly no distribution of magma intrusions.

Current geological data indicate that the three troughs (N-N-E-trending subsidence zone) that are complementary to the above uplift zones have different degrees of development in both scale and amplitude. However, generally with the exception of a few zones — the middle-southern section of the second subsidence zone — large-scale folds formed during the formation process, forming very thick Mesozoic and Cenozoic deposits.

The geological overview of the abovementioned fold zone shows that the various sections of this large tectonic system are extremely complex.

The N-N-E-trending parallel fold zones are all affected and disturbed by other tectonic systems, i.e., large E-W-trending tectonic systems, and *vice versa*. Therefore, there are various compound phenomena and joint phenomena among them, manifested in not only the overall trend of the first-level uplift zones but also the second-level and third-level tectonics in the uplift zones and subsidence zones. There are two combinations of N-N-E-trending tectonic systems and E-W-trending tectonic systems.

First, the N-N-E-trending uplift zone turns suddenly to the east or west when it approaches the E-W-trending tectonic zone, showing various degrees of arc section and forming a combined arc-shaped tectonic zone. The most prominent manifestation of this joint arc is the first first-level uplift zone in East Asia toward the Pacific Ocean. It is divided into four discontinuous sections from north to south, namely, the Kuril Islands section, the Japanese Islands section, the Ryukyu Islands section and the Chinese Taiwan–Kalimantan section. When these sections are close to the extension of the large E-W-trending tectonic zone in China, their southern ends are turned to the west, while their northern ends are turned to the east. The northern section of each island arc is also severely affected and interfered with by several N-S-trending tectonic zones, so they do not exhibit a normal N-N-E-trending pattern, i.e., the northern section of the Kuril Islands, the northern section of Honshu Island in Japan to the sea, and the Chinese Taiwan–Philippines section. Although these island arcs are affected by E-W-trending and N-S-trending tectonic systems, the straight line connecting them is generally N-N-E-trending. By the Yinshan–Tianshan large E-W-trending tectonic zone and the Qinling–Kunlun Mountain large E-W-trending tectonic zone, the third N-N-E-trending uplift zone is divided into three sections of the Daxinganling, Taihang Mountains and eastern Guizhou fold zones, which are not connected to each other. When each separated section is close to the E-W-trending tectonic zone, the southern section is slightly deflected to the SW or SWW direction, and the northern section is slightly deflected to the NE or NEE direction. However, this effect has not changed the general trend of the N-N-E-trending uplift and subsidence zones.

Second, when the N-N-E-trending fold zone crosses the E-W-trending tectonic system, there is an obvious reverse connection relationship between them in some regions. There are many examples, e.g., the middle and southern sections of the second N-N-E-trending uplift fold zone and the southern section of the third uplift zone. The N-N-E trending fold zone intersects a large E-W-trending tectonic zone in some regions, especially in the Lower Liaohe Basin north of the North China Plain and the north Henan region south of the North China Plain.[3]

The main reason for the above two compound phenomena may be related to the disappearance of the strong horizontal compression stress to form the large E-W-trending tectonic zones in the later development of the N-E-trending tectonic systems. In other words, in a certain period in the formation of the N-N-E-trending system, the strongly compressed E-W-trending system formed before the Late Cretaceous. In addition, the squeezing effect forming the N-S-trending tectonic zone is applied in the formation process of the N-N-E-trending tectonic zones, resulting in a certain section or part of the N-N-E-trending system being turned to the north or close to the normal N-S-trending direction. For example, some tectonics in the Tancheng–Lujiang fault zone, Anyang fault zone, Taihang Mountain fold zone and East Guizhou fold zone are not N-N-E-trending (18°–25° north to east) but 10°–12° north to east or even nearly N-E-trending. Additionally, because the N-N-E-trending system accommodates or merges some components of the N-E-trending tectonic systems that occurred earlier, they are arc-shaped or approximately S-shaped from north to south and even in the NE-NNE-NE direction. The southern sections of the first and second uplift zones and the vast South China Sea may be large-scale accommodation and reconnection examples.

Therefore, these large and parallel fold zones of the system are not always connected together, and the tectonic components of the system are not always N-N-E-trending.

The N-N-E-trending fold zones and compression-torsion faults in the N-N-E-trending system and the accompanying series of N-N-W-trending tension faults show large ξ-shaped tectonics in the horizontal direction. The Level II tectonics in each fold zone (uplift zone or subsidence zone) often show a ξ-shaped arrangement or geese arrangement. There are many regional ξ-shaped tectonic examples. For example, in the East Guizhou fold region, there is a S-N-trending geese-shaped or ξ-shaped tectonic staggered northeast and composed of the four Level II tectonic zones, namely, the eastern Guizhou fault zone, the Fanjing Mountain compound anticline, the Yanhe River compound anticline, and the Enshi compound anticline. In the central Hunan region, there is an S-N-trending geese-shaped or ξ-shaped tectonically staggered southwest and composed of three Cretaceous-Tertiary (Paleogene–Neogene) systems, namely, the

Yongxing–Chaling, Youxian–Liling and Hengyang–Xiangtan red basins. When looking perpendicular to the direction of these fold zones, the northwestern wings of the uplift zones are wide, while the southeastern wings are narrow. For the landform, the northwestern sections are flat, and the southeastern sections are steep. The sections with the largest thickness of new sediments or residues in troughs (deposit zones) are generally close to the western side of the subsidence fold zones. This feature is extremely obvious in the cross sections of Daxinganling, Songliao Basin, Zhangguangcailing, Xihotaling Mountain, Sea of Japan, and Japanese Islands, the cross sections of Shaanxi–Gansu–Ningxia Basin, Taihang Mountains, North China Plain, East Shandong, Yellow Sea and Ryukyu Islands, and the cross sections of Sichuan Basin, East Guizhou Fold Zone, central Hunan Basin, Wuyi Mountain to Daiyun Mountain, Taiwan Strait and Taiwan Province. The shape of these parallel large fold zones is extremely asymmetrical, but a common feature is that in the uplift zones, the eastern wing is steep while the western wing is gentle. In contrast, in the subsidence zones, the western wing is steep, and the eastern wing is gentle. On the steep side of the uplift zones, there are often the same-trending large torsional faults or fault zones dipped westward, i.e., the deep trough on the eastern side of the first level I uplift zone, the Nenjiang–Qihar fault (or fault zone) on the eastern margin of Daxinganling on the northern section, the East Taihangshan fault in the middle section, and the East Guizhou fault on the southern section of the third uplift zone. From a general point of view, the axial planes of these asymmetrical, mutually parallel, and nearly N-W-W-trending large fold zones, and the large fault surfaces form an imbricated tectonic. Over a wide range, the N-N-E-trending tectonic systems clearly show not only a ξ-shaped form on the plane but also a ζ-shaped form on the section, which is an extremely important feature in the development of these tectonic systems.

During the development of the N-N-E-trending systems, frequent and intense igneous rock activity occurred. They are mainly acidic intrusive rocks and acidic and intermediate-acid extruded rocks and in some regions, small ultrabasic rock masses or dykes. The granite has an extremely wide distribution, and the scale and quantity of the rock masses or rock zones are comparable to those of the granite in certain sections controlled by the large E-W-trending tectonic zones. Seen from the surface, the granite masses related to the N-N-E-trending tectonic formations are concentrated mainly in the second uplift zone, especially in its southern section and the Daxinganling region on the northern section of the third uplift zone. In some sections of these regions, the granites are exposed in a scattered pattern, and the rock mass trend is not obvious, but most of the rock masses or rock zones are distributed in the same direction as the local N-N-E-trending system. In tectonic regions compounded with large E-W-trending tectonic zones, the spatial relationship between the distribution of granites and the N-N-E-trending tectonic systems is still obvious and is easy to distinguish. These rock masses and rock zones have been shown to be strongly controlled by various levels of tectonics in the N-N-E-trending system.[4] In the southern section of the second uplift zone, there are five large granite zones intermittently exposed in the Jiangxi, Guangdong, and Fujian regions. From east to west, they are the Southeast Coastal Margin zone, Haifeng–Nanjing–Daiyun Mountain–Pingnan zone, Gutian–Shunchang–Pucheng zone, Heyuan–Wuping–Guangze zone and Yudu–Yihuang zone. On the western wing of the northern section of the second uplift zone of the N-N-E-trending tectonic system, the distribution of granites is also obvious in Jilin, including the Wangqing–Yanji zone on the east and the Binxian–Panshi zone on the westernmost margin. In the northern section of the third uplift zone of the N-N-E-trending tectonic system, there are the Right Front Horqin Banner–Wenduhada zone to the south and the Nenjiang River zone and the Jiliu River zone to the north of Daxinganling, as well as the Xiaowutaishan granite zone north of the Taihang Mountains in the middle section. In addition, in some regions of the large E-W-trending tectonic zone that divides the N-N-E-trending subsidence zones, the granites are also widely distributed. In the Yanshan region, there are the Fuxin–Jinzhou–Funing zone, Chifeng–Chengde zone and Yanqing–Fengning zone from east to west. In the Nanling region, there is the Yangjiang–Zhaoqing–Sihui zone. Notably, in some sections, although many granites appear as isolated small-scale rock masses on the surface, they are actually often connected along the underground tectonic trend, which is very important in terms of prospecting.

Another manifestation of igneous rock activity during the formation of this tectonic system is the massive eruption of acidic and intermediate-acid volcanic rocks. These volcanic rocks appear mainly in the southern section of the second uplift zone, namely, the vast region of the southeastern coast of China and the entire Daxinganling region in the northern section of the third uplift zone. It is a large-scale N-N-E-trending uplift zone. For these volcanic rocks, it is difficult to distinguish the number of eruptions in succession, but the two eruptions in the Late Jurassic and Early Cretaceous are the most important. More details will be described when explaining the formation period of the tectonic system.

The internal tectonics of these large fold zones (uplifts and folds) are quite complex. Generally, each fold zone is often composed of a number of Level II, Level II or lower-level tectonic zones with different sizes and shapes. These lower-level tectonic components are mainly compound or single anticlines and synclines, as well as more complex torsional and compression faults and fault zones, followed by a large number of tensile and torsional faults associated with these compressional tectonics. In the normal distribution, the tensional faults are nearly orthogonal to the N-N-E-trending fold axis; that is, they are N-W-W-trending. The torsion faults obliquely intersect the N-N-E-trending fold axis; that is, they are N-E-E-trending or N-N-W-trending, i.e., the Taishan fault and the Dayishan fault. In addition, there are often various torsion tectonics with different scales and shapes in tectonic zones. On the side of certain faults with larger twists, oblique folds and compressional faults are often derived. Both of them together form a herringbone tectonic. In the entire eastern region, the development of associated tectonics varies from place to place. In many regions, two sets of torsion faults often form excellent checkerboard tectonics. In some regions, they are few and scattered, and only one set of faults even exists.[5]

5.1.2 *Uplift Zones*

(1) First Level I Uplift Zone
This zone is the easternmost of the N-N-E-trending large uplift zones and Taiwan Province is located in its central-southern section. In Taiwan Province, this uplift fold zone is manifested mainly as the asymmetric Taiwan Mountain compound anticline with a steep eastern wing and gentle western wing composed of Paleozoic strata and Paleogene–Neogene strata and compression and torsional faults parallel to this anticline. Due to the strong influence of the Nanling E-W-trending tectonic zones and the N-S-trending tectonic zones, the northern section of the Taiwan Mountain compound anticline is obviously bent toward the northeast, while its southern section is nearly turned to the normal southern direction, instead of being N-N-E-trending from beginning to end.

(2) Second Level I Uplift Zone

This zone is in the middle of the three large uplift zones. Based on their distribution features in China, this zone can be divided into the Liaoning–Jilin–Heilongjiang East Slope–Shandong Peninsula zone and the Wuyi Mountain–Daiyun Mountain zone.

(a) *Liaoning–Jilin–Heilongjiang East Slope–Shandong Peninsula zone*: Obviously, the mountains east of Heilongjiang, Jilin and Liaoning Provinces and the vast region of the Shandong Peninsula east of the Yishu fault zone are only the western wing of the N-N-E-trending uplift fold zone, not the whole zone, which includes the northern section. In this region, due to the large-area exposure of former Hercynian granites and relatively few depositional strata, the Levels II and III tectonics are mainly N-N-E-trending torsional and compressional faults or fault zones and a large number of granite masses rather than large-scale complete folds. First, there are two Cretaceous N-N-E-trending synclines clearly staggered north to south to southwest in northwestern Yichun. Second, on the western margin of Liaoning, Jilin and Heilongjiang, approximately from west of Yilan to northwest of Yitong, in the Hercynian granite and Jurassic strata, the tectonics are manifested as a series of Yanshanian granite masses. Although they are not connected into a zone, each rock mass is N-N-E-trending. From south of Changchun–Jilin to the Shandong and Anhui regions, the large-scale faults and fault zones are concentrated. Although these tectonics extend for a long distance, passing through the Yinshan E-W-trending tectonic zone, Qinling E-W-trending tectonic zone, and the vast Bohai Sea region, they are basically N-N-E-trending, except that the N-N-E-trending is turned to the north in the Shandong region. Based on their distribution from the northeastern region to the Shandong region, they can be approximately divided into three parallel tectonic zones from east to west, namely, the Liuhe–Fengcheng fault zone, the Siping–Yingkou–Jinxian fault zone and the Yishu fault zone.

Liuhe–Fengcheng fault zone: This zone's northernmost section starts from the western side of the Liuhe Basin, and the middle section is relatively fuzzy but quite obvious in the Fengcheng region. This tectonic zone basically extends at 35°E from north to east. It intersects not only the large E-W-trending tectonic zone but also the N-E-trending tectonic zone. Different rock formations are cut in different locations, including the pre-Sinian metamorphic rock series system and ancient granites and the Jurassic system in many sections.

Siping–Yingkou–Jinxian fault zone: The northern section of this zone starts from east of Siping and extends southward through Liaoyang, Yingkou, Xiongyue, Bohai Sea, Penglai, Laiyang and southern Rizhao in the Shandong Peninsula, and its trace is still very obvious. Obviously, its northern section is close to the western margin of the second uplift zone. From an overall point of view, the northern section extends at approximately 35° north to east, and the southern section generally extends at 5°–30° north to east. They mostly intersect the pre-Sinian and Paleozoic strata in the Liaodong Peninsula. The latest strata are Jurassic–Cretaceous and Yanshanian granites in the Shandong Peninsula.[6] This tectonic zone intersects the Yanshan E-W-trending tectonic zone, and the N-E-trending tectonic zone is more intense than the previous fault tectonic zone.

Yishu fault zone: This zone extends in the central part of Shandong on an extremely large scale, with a very obvious trace and very large width. It controls the direction of Yishui and Shushui. Due to the strong influence of the Yishu fault, the pre-Sinian subsystem is directly disconnected from the Cretaceous system in this zone, and the rock formations are very fragmented. Based on geophysical prospecting work in the Bohai Sea, the traces of this fault zone are still quite obvious in Laizhou Bay, and the zone may be integrated with the previous fault zone on the eastern margin of the lower Liaohe trough. Aeromagnetic data indicate that this tectonic zone still exists on a large scale in the Jiangsu and Anhui regions. This tectonic zone starts from Tancheng in southern Shandong, passes through Xinyi, Suqian, and Sihong to the Lujiang south of Jiashan Mountain and is integrated or intersected obliquely by the eastern wing of the Huaiyang epsilon-shaped tectonic zone. At the same time, it also clearly intersects the northern branch of the Qinling E-W-trending tectonic zone — the Songshan–Tongxu–East China Sea compound anticlines on the southern margin of Shandong. In general, similar to that of the previous two zones, this fault zone is still N-N-E-trending to the east, but the southern section is slightly N-N-E-trending to the north. Considering the drastic changes in aeromagnetic anomalies and the occurrence of more than one major earthquake in recent years, it has been suggested that it is a deep fault zone in the crust and that there may be a large number of basic or ultrabasic rock masses in deep locations.

(b) *Wuyi Mountain–Daiyun Mountain zone*: This uplift zone is composed mainly of Wuyi Mountain and Daiyun Mountain. It starts in northern Zhejiang, passes through western Fujian and southeastern Jiangxi, and reaches eastern Guangdong to the south, with an obvious presentation in southeastern China. The Pre-Sinian Banxi Group and Jianping Group are widely exposed in Longquan, Chong'an, Jiangle, Nanping, Sanming and other places, forming the large axis of the uplift zone. The upper Jurassic–Cretaceous intermediate acid volcanic rocks are generally distributed in the eastern coastal region, forming the eastern wing of the uplift zone, while the Lower Paleozoic strata are distributed on the western wing in eastern Jiangxi. This uplift zone has extremely significant tectonic traces and is composed of large-scale compound anticlines, large fault zones, and a large number of Yanshanian granites associated with these tectonics. These tectonics are generally N-N-E-trending. However, when they are connected to the Nanling large E-W-trending tectonic zone, although the compound anticline folds become so weak and even disappear, some large faults turn in the S-W-trending or S-W-W-trending direction, forming an arc that protrudes slightly to the southeast. This characteristic is different from the behaviors of the northern section of the uplift zone crossing the E-W-trending zone. The Wuyi Mountain–Daiyun Mountain uplift zone can be approximately divided into six Level II tectonic zones from east to west, namely, the southeast coastal tectonic zone, the Lishui fault zone, the Daiyun Mountain compound anticline, the Jiangle compound anticline, the Wuyi Mountain compound anticline and the Nancheng–Yudu fault zone.

Southeast coastal tectonic zone: This zone is composed mainly of torsional compression faults, compression zones and intermittently distributed Yanshanian granites. It extends at 20°–30° north to east to the north of Putian and gradually turns from 30° to 55° north to east to the south of Putian. Undoubtedly, some sections of its eastern wing are embedded in the sea. The tectonic zone has a maximum width of hundreds of meters.

Lishui fault zone: This zone starts in Zhuji in northern Zhejiang Province, passes through Lishui, and ends in the Shouning region. It is the smallest among these six zones. It is composed mainly of three large faults arranged diagonally parallel to each other. Except for the interference of the N-S-trending tectonic zone in its southern section, the entire fault zone extends in the N-N-E-trending direction. Due to strong compression, there is strong fragmentation in the Late Jurassic–Cretaceous intermediate-acid volcanic rocks in this region.

Daiyun Mountain compound anticline: This anticline is actually a continuous tectonic zone composed of the Daiyun Mountain great anticline, the Zhangping great fault, and the east Guangdong Jiexi great fault. It is the longest and most obvious among these parallel tectonic zones. It starts from the northwestern region of Jinhua of Zhejiang in the north, passes through the Nanling E-W-trending tectonic zone in the south, enters eastern Guangdong, and is then embedded in the South China Sea. The N-N-E-trending pre-Sinian system is intermittently distributed along the axis of the Daiyun Mountain compound anticline, and there are mainly Jurassic and Cretaceous intermediate-acid volcanic rocks on its eastern and western wings.[7] Due to the intersection of the Sanming Great Fault in the same direction, the compound anticline is extremely narrow on its western wing. The development features of this tectonic zone are as follows: (1) The Daiyun Mountain N-N-E-trending large compound anticline disappears suddenly when it approaches the Nanling E-W-trending tectonic zone. (2) The Zhangping fault and the Jiexi faults occur one after another and turn from the original N-N-E-trending direction to the S-W-trending and then the S-W-W-trending direction.

Jiangle compound anticline: This anticline starts from eastern Chong'an to the north, passes through Jiangle to the south, and finally reaches the Gutian region, which is also a large N-N-E-trending anticline. Its northern section and middle axis are composed of the pre-Sinian and Lower Paleozoic systems, and the Upper Paleozoic and Jurassic systems are generally distributed on its two wings. The axis of its southern section is the Yanshanian N-N-E-trending Gutian granite masses, and its two wings are also composed of the Upper Paleozoic and Jurassic systems. The Jianyang N-E-trending compound anticline is intersected obliquely by the middle section of the Jiangle compound anticline, while the Nanling E-W-trending tectonic zone is reversely intersected by it and finally disappears on the eastern side of the spine of the Meixian epsilon-shaped tectonic zone in Guangdong.

Wuyi Mountain compound anticline: This anticline is one of the two compound anticlines on the southern section of the second uplift zone of the system. The general trends of the scale, morphological features and distribution of its development location are very similar to those of the Daiyun Mountain compound anticline on its east, which is also formed by anticline folds and a large fault. The southern section of the anticline starts from Wuping, passes through eastern Nanfeng to the north and reaches the Zixi region. There is a fault zone dominated by the Heyuan faults south of Wuping. The oldest Lower Paleozoic strata appear on the axis of the Wuyi Mountain anticline, and its eastern and western wings are covered by Jurassic and Cretaceous N-N-E-trending systems. The most obvious traces on the southern section compounded with the E-W-trending Nanling tectonic zone are the faults and the Yanshanian granite zone. Only the Heyuan fault and the granite zone extend farther, and the N-N-E-trending direction gradually turns to the N-W-trending or N-W-W-trending direction. Most of the remaining faults basically disappear on the southern margin of the Nanling E-W-trending tectonic zone.

Nancheng–Yudu fault zone: This zone is composed of many N-N-E-trending faults and a series of N-N-E-trending Yanshanian granite zones with obvious traces in southeastern Jiangxi. The eastern and western sections of this zone are completely different in terms of tectonic form and stratigraphic distribution. This zone is a tectonic zone with extremely strong changes in the western boundary of the southern section of the second new Cathaysia uplift zone. This description becomes increasingly obvious southeast of Linchuan, and the early Paleozoic and Jurassic rock masses intersected with it are subject to violent compression and fragmentation. The earlier N-E-trending systems intersect obliquely by their northern and middle sections, which are intersected by the later N-E-trending tectonic zones. They are strongly interspersed in the Nanling E-W-trending tectonic zone in the southern section and intersect reversely by a series of E-W-trending faults, forming a grid structure. However, these faults are always N-N-E-trending in the Nanling E-W-trending tectonic zone but are curved to the west, so they seem to be against the N-W-W-trending arc-shaped tectonic zone to the east.

(c) *Third Level I Uplift zone*: This zone is a tectonic zone with the largest width, the most spectacular shape, the best continuity, and the most obvious reflection of various features among the three N-N-E-trending large uplift and fold zones. Due to the separation of the Yinshan–Tianshan and Qinling–Kunlun E-W-trending tectonic zones, they are divided into three major sections from north to south, namely, the Daxinganling, Taihang

Mountains and East Guizhou Fold Zones, which are staggered westward from north to south, together with the subsidence zone to the east. However, their main bodies are uniformly N-N-E-trending.

Daxinganling region in the northern section of the uplift zone: This uplift zone features the widespread distribution of Yanshanian acidic and inter-mediate-acid volcanic rock systems, as well as intermittently exposed Paleozoic or Upper Paleozoic systems and a large number of Hercynian granite masses. Their distribution is N-N-E-trending and is controlled by the Level II tectonics of this uplift zone, which is specifically reflected as a N-N-E-trending compound anticline and two N-N-E-trending com-pound synclines, namely, the east Dayangshu compound syncline, the middle Oroqen compound anticline and the west Keyihe compound syn-cline. On the eastern margin of the uplift zone, there is a nearly N-N-E-trending large torsional fault zone along the Nenjiang River. The eastern Dayangshu syncline is actually composed of the Dayangshu syncline, the Yadong syncline, the Ulanhot trough and the Bahrain syncline from north to south, and the strata controlled by all sections are composed mainly of Jurassic–Cretaceous volcanic rocks.[8] The northern and middle sections of this large syncline are basically N-N-E-trending, but when the southern section is turned to S-W-trending and S-W-W-trending approximately in the western River region, forming a prominently convex arc to the east. The middle Oroqen compound anticline is undoubtedly the main tectonic component of Daxinganling in terms of its scale. Its northern end starts approximately from the Huma'er River, passes through the Jiagedaqi, Buteha Banner and Hulin Rivers to the south, and finally reaches the Xilamulun River. Its axis is composed of the large Oroqen–Buteha Banner Hercynian granites in the northern section and the oldest stratum, namely, the Upper Paleozoic metamorphic rock system exposed in the middle and south sections. Its two wings are composed of the Jurassic–Cretaceous volcanic rock system from north to south. Although the northern section is intersected by E-W-trending tectonic zones and the middle section is disturbed by a series of N-E-trending tectonic zones, it still passes through Daxinganling and spreads obviously N-N-E-trending from north to south. The southern section starts from the Harin River and reaches the upper reaches of the Xilamulun River. Similar to the syncline zone on its east, it turns sharply to S-W-trending and S-W-W-trending, showing a large arc protruding to the southeast. West of this anticline, there is an extremely wide N-N-E-trending Keyihe syncline. It starts from the Keyihe River to the north and extends to the south, disappearing in the Suge River. The main strata controlled by this great syncline are composed of Cretaceous volcanic rocks.

Taihang Mountain region in the middle of the uplift zone: Because it is overlapped and compounded on the "Shanxi block", this uplift zone should actually include the entire "Shanxi block", not just the Taihang Mountain part. The internal tectonics of the Taihang Mountains include not only N-N-E-trending compounds and single folds but also large-scale torsional fault zones. These tectonics are more concentratedly divided into two zones: the eastern zone and the western zone. The eastern zone is the main body of this uplift zone, which strongly controls the eastern margin of the "Shanxi block" and forms the so-called "Taihang uplift". In the northern section of the "Taihang Uplift", there is a large Fuping compound anticline. Its axis is composed of the Hutuo system and Wutai system on the pre-Sinian boundary, and its eastern and western wings are composed of the widely distributed Sinian subsystem, Cambrian system, Ordovician system and other strata. In the southern section of the "Taihang Uplift", there are the Zanhuang compound anticline and the Gaoping–Xiong'ershan anticline, and they are generally composed of the pre-Sinian and Paleozoic systems. Between these two compound anticlines, there is an extremely wide Qinshui syncline, which is a large red basin dominated by the Triassic system. Obviously, on the eastern margin of the "Shanxi Block", these N-N-E-trending folds are arranged diagonally parallel to each other from north to south, forming a magnificent geese-shaped ξ-shaped tectonic zone.[9] However, some sections are also greatly disturbed by the N-S-trending tectonic zones. On the other hand, due to the interference of the Qinling Mountains E-W-trending zone, the N-N-E-trending tectonic zone turns from the normal N-N-E-trending to S-W-trending and S-W-W-trending. For example, in the case of the Gaoping–Xiong'ershan compound anticline, its northern section is basically N-N-E-trending, extends to the southwest, is affected by the southeast Shanxi epsilon-shaped tectonics, and is severely covered by the Quaternary system in the Jiyuan–Luoyang region. Then, in southwestern Luoyang, a large compound anticline appears; however, this section is not N-N-E-trending, but S-W-W-trending. Therefore, this anticline zone forms a joint arc protruding significantly to the southeast from north to south. There is a tectonic zone on the western side of the "Shanxi Block"; one part appears approximately on the upper reaches of the Fen River, and the other part is distributed in the southeast of the lower reaches of the Fen River, approximately along Taiyue Mountain to the eastern side of Zhongtiao Mountain. The former is composed mainly of the Jurassic and Cretaceous Datong syncline and a series of folds in the Jingle region to its south, while the latter is composed mainly of fault zones. Obviously, they are not closely connected to each other but are arranged diagonally from north to south to southwest in the N-N-E-trending direction. This fault zone on the southeastern side of the Fenhe River passes through the Yellow River and then forms a large Yaoshan compound anticline. Similar to Xiong'er Mountain in the east, it turns sharply from N-N-E-trending to S-W-W-trending, forming another large joint arc tectonic between the "Shanxi Land Platform" and the Qinling Mountains large E-W-trending zone in the west. The relationship between the northern section of Taihang Mountain and the Yanshan Mountain E-W-trending zone is more complicated, with compound and overlapping features.

East Guizhou fold zone region in the southern section of the uplift zone: The geological features in this uplift region are quite different from those in its northern and middle sections. Its most prominent feature is the lack of igneous rocks related to the generation of the N-N-E-trending tectonic system, though the internal tectonics are extremely well developed. Not only are there large N-N-E-trending fault zones, but multiple folds parallel to the fault zones are also distributed greatly. From the eastern Guizhou fault to the western Huayingshan anticline zone, this region can be divided into at least five N-N-E-trending tectonic zones, namely, the East Guizhou fault zone, Fanjingshan compound anticline, Yanhe compound anticline, Enshi compound anticline and Fangdoushan–Huayingshan fold group. These tectonic zones are generally accompanied by narrow synclines. The composed strata or those affected by these tectonics include those formed in various periods before the Middle Triassic. The axis of the compound anticline is generally composed of the Pre-Sinian Banxi Group or the lower part of the Lower Paleozoic system, and the two wings are composed mostly of the upper part of the Lower Paleozoic system.[10] These tectonics are not only extremely parallel to each other but also obviously staggered from north to south in the W-S-trending direction, showing an extremely perfect geese-like ξ-shaped tectonic zone. In Liuzhou, Yishan, Hechi and other regions northwest of Guangxi, these tectonic zones clearly intersect the Nanling E-W-trending tectonic zone. They also do not cross the entire Nanling E-W-trending tectonic zone but gradually disappear. The various tectonic zones extend northward along this uplift zone and obliquely intersect the N-E-trending and N-E-E-trending tectonic zones, and the northern section also shows a slight eastward bending trend. In addition, some of the tectonic zones are obviously affected by the Sichuan–Guizhou N-S-trending tectonic zones; for example, the Yanhe and Enshi compound anticlines are obviously intersected. Because of this interference, the N-N-E-trending tectonic zones in the Zhenyuan region of Guizhou are significantly turned to the south, with accommodation or transition relationships between them.

5.1.3 *Depression Zones*

The data show that the three subsidence zones accompanied by the three N-N-E-trending large uplift zones are not all large and simple depressions, and there are some obvious differences among these zones or different sections of a certain zone. In some zones or certain sections of a zone, the internal tectonics are quite complicated, and there are often several uplift (multiple or single anticlines) fault zones and multiple or single synclines with different sizes and shapes. Some sections of these Level II or III nearly N-N-E-trending tectonic zones show distribution features and combination patterns that are very similar to the internal tectonics of the uplift zone.[11] At present, the depression zone cannot be fully described, and only some of the most prominent sections are explained (Figs. 5.2 and 5.3).

The second depression zone is the most complicated among the three depression fold zones, and the subsidence in the southern section is slightly smaller than that in the northern and middle sections, which is a great development feature of this zone. There are significant Level II tectonic zones in the northern Songliao Plain, the central North China Plain and the Qinling–South Dabie Mountains Plain, which are increasingly complex from north to south. Geophysical data show that in the Songliao Plain and the North China Plain, the Level II tectonic zones in some sections are composed of compound or single anticlines, as well as faults or fault zones. They significantly intersect obliquely or reversely with the N-E-trending tectonics and compound and intersect the N-E-trending tectonics. In the Yanshan Mountain E-W-trending tectonic zone between these two plains, the Level II tectonics are particularly obvious and concentrated between Jinzhou and Chifeng. From east to west, they include the Jinzhou anticline, Liujiazi syncline and Jianping–Xifengkou

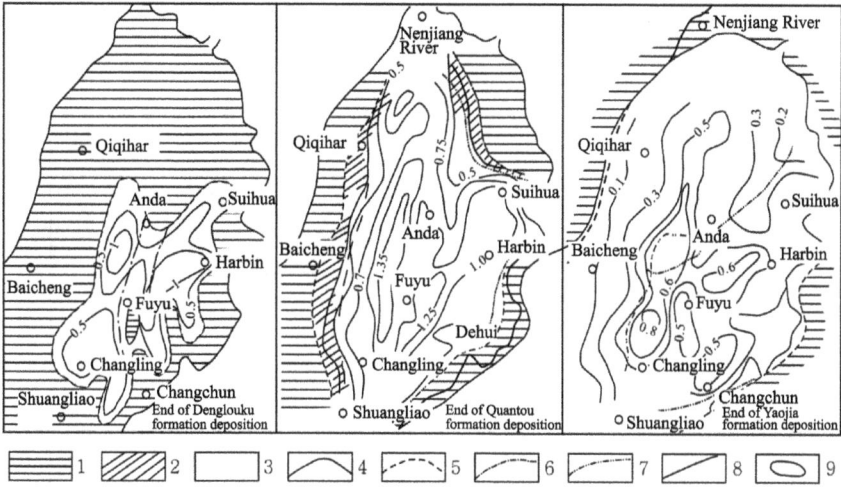

Fig. 5.2. Schematic diagram of paleotectonism and paleogeography in different Mesozoic and Cenozoic periods in the Songliao Basin (Unit: m).

Notes: (1) Uplift and denudation zone; (2) Lakeside region; (3) Depression region; (4) Modern basin boundary; (5) Uplift and depression boundary; (6) Tectonic zone boundary; (7) Paleogeographic zone boundary; (8) Fault; (9) Deposition contour line.

large fault. The strong E-W-trending tectonics and some N-E-trending tectonics are obviously intersected by them in the region. These Level II tectonic zones do not correspond with and are not continuous with the same tectonic zones on the northern and southern sides of the plains.

South of the Qinling–Dabie Mountains, Level II tectonic zones are widely developed in the vast region between eastern Hunan and western Jiangxi, with extremely significant traces and considerable scales. They start from the Dabie Mountains to the north, pass through eastern Hunan, western Jiangxi and central Guangdong to the south, and finally are submerged in the South China Sea. This tectonic zone is composed mainly of a series of N-N-E-trending anticlines, large-scale faults and fault zones and several Cretaceous-Tertiary (Paleogene–Neogene) red basins parallel to them. Based on its distribution features, it can be roughly divided into a northern section and a southern section.[12] In the section north of the Jiuling Mountains and south of the Dabie Mountains, the N-N-E-trending faults are relatively scattered, but their trends are extremely consistent. East of the Qinling Mountains, it intersects not only the southern branch of the Qinling Mountains E-W-trending zone but also the front arc of the

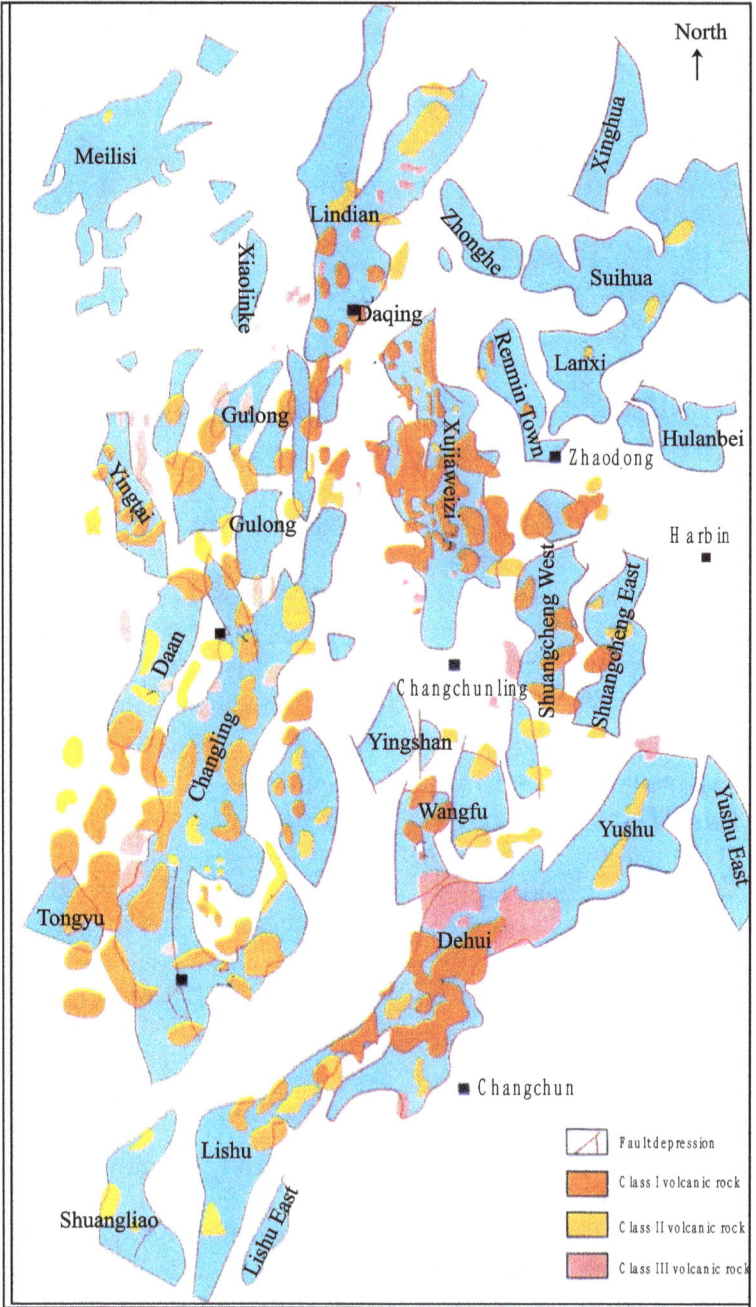

Fig. 5.3.   Distribution of Mesozoic fault depressions and Mesozoic volcanic rocks in the Songliao Basin.

Huaiyang epsilon-shaped tectonic zone and the N-E-trending tectonic zone. The southern section is divided into an eastern zone and a western zone. The eastern zone spreads between eastern Hunan and western Jiangxi, and its main body is a N-N-E-trending mountain composed of Luoxiao Mountain, Wanyang Mountain and Zhuguang Mountain, together with a N-N-E-trending large torsional fault. This zone has very obvious geomorphology. It is overlapped and compounded on a large N-S-trending uplift tectonic zone in this region, so there is interference from the N-S-trending tectonics. This zone extends along Shaoguan, Yingde and finally reaches the Yangjiang region of Guangdong, significantly passes through the Nanling E-W-trending tectonic zone, gradually turning from the original S-S-W-trending to the S-W-trending. The western zone spreads in the eastern Hunan region, and its main body is composed of the famous red basins and large faults parallel to the basins in eastern Hunan. These basins include the Yongxing–Chaling Basin, Youxian–Liling Basin and Hengyang–Xiangtan Basin. They are not only distributed parallelly in the N-N-E-trending direction but also prominently staggered from north to south in the southwest direction, forming medium-scale geese-like ξ-shaped tectonics in this region. The western zone extends southward, forming the powerful Dayunwu Mountain fault zone in southwestern Guangdong. Similar to the eastern zone, it intersects the E-W-trending zone in Nanling and then turns from the S-S-W-trending to the S-W-trending zone. In addition, for the southern section of the second subsidence zone, Level II tectonic zones are also obviously distributed east of Guangxi and concentrated in northwest and southeast of Darong Mountain. Their distribution features are very similar to those of the broad eastern zone in Guangdong, and they also turn from S-S-W-trending to S-W-trending.

The Level II tectonic zone in the third depression zone is located mainly in southwestern Sichuan and eastern Yunnan. It starts in the middle of Longmen Mountain in western Sichuan to the north, passes through Ya'an, Leshan, Leibo, Qiaojia, Dongchuan, Qujing and Kunming to the south, ending at Gejiu. It does not intersect the Ailao Mountain ξ-shaped tectonic zone in the Qinghai–Tibet–Sichuan region and is not affected by this tectonic zone. Instead, it gradually disappears near the Ailao Mountain tectonic zone. This tectonic zone is a N-N-E-trending large compressional fault or fault zone accompanied by a few small-scale folds that coincide with each other. From the perspective of its distribution, it intersects a variety of tectonic systems and is also interspersed by other tectonic systems in some regions, but the former feature is dominant and obvious. In the northern section of this tectonic zone, the Ya'an and Leshan tectonic zones are concentrated in southwestern Sichuan. In the northern section of the Ya'an tectonic zone, a N-N-E-trending large fault clearly intersects the eastern wing of the N-E-trending system and the Jinchuan arc tectonic zone and is intersected by a newer N-E-trending tectonic zone. In the southern section of the Ya'an tectonic zone, the Ya'an anticline obliquely intersects the N-W-trending ξ-shaped tectonic zone at a clear angle.[13] The Leshan tectonic zone is composed mainly of a N-N-E-trending anticline and two parallel faults. It starts from eastern Chengdu and extends to the south of Leshan; it is a tectonic zone with remarkable features along the southwestern margin of the Sichuan Basin. The Jurassic and Cretaceous strata intersected by some tectonic zones show strong compression and fragmentation. In the middle of the tectonic zone between Zhaojue of Sichuan and Zhaotong of Yunnan, there are only a few N-N-E-trending fault tectonics, which intersect the Upper Paleozoic Mesozoic Jurassic–Cretaceous strata in this region. In the vast eastern Yunnan region south of Dongchuan and Guizhou in the southern section of the tectonic zone, the traces are very obvious, and they are N-N-E-trending large compressional faults and fault zones. The Paleozoic system and part of the Mesozoic system in this region have been severely affected and faulted by staggery. They are compounded with the Yunnan epsilon-shaped tectonics and the Sichuan–Yunnan N-S-trending tectonic zones, especially in the vast eastern Yunnan region, and it is difficult to distinguish the epsilon-shaped eastern wing in a small or partial region. Judging from the general trend of the extension of this section of the tectonic zone, its trend to the east is slightly lower than that of the tectonic axis of the eastern wing of the Yunnan epsilon-shaped tectonic zone. It is compounded reversely with the epsilon-shaped tectonic zone on the eastern wing reflection arc and arc top section. The above situations show that the faults and folds of the three N-N-E-trending tectonic zones are not only consistent and parallel but also obviously staggered from north to south in the southwest direction, forming an extremely rare geese-shaped ξ-shaped tectonic in a very prominent, large-scale and new large N-N-E-trending Level II tectonic zone.

In this region, the development of the Level II tectonic zones is very different from that inside the second fold zone. On the one hand, the uplift shape is not prominent. On the other hand, the anticlines and faults and the N-N-E-trending tectonics on the western margin of the fold zones together form ξ-shaped tectonics, that is, the Level II tectonics and the western margin in the fold zone are gradually transited.

5.1.4 *Genetic Relations and Development Period of Tectonic Systems*

A large amount of data has shown that the N-N-E-trending tectonic system was generated in the Late Carboniferous and reached a peak in the Late Mesozoic (Cretaceous) and Cenozoic, which was still active in some sections from the aspects of deposits, tectonic relationships or igneous rock activities. From the point of view of the tectonic system, the first and second fold zones of the N-N-E-trending tectonic system have long been regarded as large depression zones that have gradually developed since the Mesozoic, which has been proven by the results of a large number of geological and geophysical works in the past 20 years, especially the past 10 years. These two depression zones not only generally control the Paleogene–Neogene and Cretaceous deposition strata with a large thickness but also deposit the Jurassic system in some sections with large folds. These two depression zones, similar to the third depression zone in the west, were generated before the Cretaceous, and the Jurassic system was developed completely and widely distributed in the east (the western wing of the northern section of the second uplift zone) and west (the northern section of the third uplift zone) of the Songliao Plain and the Yanshan region in the south. The strong tectonic unconformities of the middle–upper Jurassic system can be seen in Tonghua, Fuxin, Beipiao, Hulin River, Mosquito River and Zhalai Nuoer, which control the N-E-trending middle-lower Jurassic tectonics and N-N-E-trending upper Jurassic tectonics. These N-E-trending folds are related to the N-E-trending tectonic system. The N-N-E-trending upper Jurassic folds are undoubtedly low-level tectonic components that reflect the first-level uplift zones and the first-level fold zones of the N-N-E-trending tectonic system and strongly indicate that the earliest geological generation period of the N-N-E-trending first-level large uplift was not the Early Jurassic but the early Late Jurassic.

South of the Jiuling Mountains in Jiangxi in the Jingdezhen region, the N-E-E-trending or N-E-trending tectonics composed of the middle-lower Jurassic and the Carboniferous–Permian systems or Triassic system are more obvious. Although the N-NE-trending upper Jurassic system has not been found to be directly unconformable on these tectonics in these regions, it is obviously transected by the nearby N-N-E-trending Lower Cretaceous tectonics (new syncline tectonics), which also indicates that the earliest N-N-E-trending tectonic system in this region is not earlier than the Late Jurassic.

North of Jiangle–Nanping in the Fujian region, the upper Jurassic N-N-E-trending syncline becomes N-E-trending in the south, and the N-E-trending syncline is composed of the Middle-Lower Triassic system and the unconformable Middle and Lower Jurassic system. It seems that the system adapted to the earlier N-E-E-trending tectonics, which also indicates that the N-N-E-trending tectonic system was generated in the Late Jurassic.

The third fold zone is different from the first two-fold zones in that the Mesozoic strata are well developed, and the Paleogene–Neogene deposits are rare and more complicated. Due to the strong influence of E-W-trending compression after the Cretaceous or Oligocene, the Ordos Basin is not N-N-E-trending but nearly N-S-trending.[14] The Sichuan Basin south of the Qinling Mountains is not N-N-E-trending but normal N-E-trending. However, the connecting line between the centers of these two basins is neither N-S-trending nor N-E-trending but N-N-E-trending, which is extremely consistent with that of the third uplift zone in the east. Therefore, these two large Mesozoic basins are controlled by the N-N-E-trending fold zones, forming a large fold zone in the New Cathaysia fold zone. However, it was not controlled by this system at the beginning of the Mesozoic, nor was it generated earlier than the other two-fold zones. Obviously, in the tectonic system, the Ordos Basin is not only a compo-nent of the N-N-E-trending fold zone but also the eastern shield of the Qilu–Helan epsilon-shaped tectonic zone, which shows that in the geo-logical development process, it is controlled by not only the N-N-E-trending system but also the epsilon-shaped tectonic system. Because the Middle and Lower Jurassic system is generally controlled by the epsilon-shaped two wings and ξ-shaped trough, sediments before the Late Jurassic in the Ordos Basin can reflect the control of the epsilon-shaped tectonics. Sediments after the Middle Jurassic can be controlled only by the system's trough and overlapped and compounded on the shield controlled by the epsilon-shaped deposits. This relationship is also extremely prominent in the Baihua Mountain region west of Beijing. Here, the middle-lower Jurassic Qilu–Helan N-E-E-trending epsilon-shaped east-wing syncline is not integrated by the N-N-E-trending upper Jurassic system, and both are obviously compounded reversely.

The large Longmen Mountain tectonic zone on the western margin of the Sichuan Basin is an extremely complex zone that has undergone repeated drastic changes. After the Cretaceous, the N-E-trending and N-E-E-trending tectonics generated by the large E-W-trending torsion may reflect the last drastic change. This change may have played a major role in the reformation of the formed basin contours, but there is little or no obvious relationship between it and the generation and development of the Sichuan Basin. In recent years, geological works in this region have found that there was a very violent tectonic change between the Xujiahe Formation (possibly the Upper Triassic system) and the Baitianba Formation (possibly the lower Jurassic system) in the Upper Triassic or Triassic–Jurassic system on the eastern margin of the tectonic zone. The large anticline that turns to the basin below the unconformity is N-E-trending, and the middle-lower Jurassic system above the unconformity is gently overlaid from southeast to northwest (that is, perpendicular to the fold axis). Obviously, the N-E-trending middle-lower Jurassic systems controlled by the N-E-trending tectonics are completely different from the N-N-E-trending Ya'an and Leshan tectonic zones in this region. The former cannot also be controlled by the N-N-E-trending system. Based on the distribution features of the tectonic systems in East China, the middle-lower Jurassic system reflects or is controlled by the N-E-trending tectonic zones rather than by the system. Therefore, as a N-N-E-trending trough, the Sichuan Basin was developed and overlaid only from the Middle Jurassic on a large N-E-trending basin deposited for a long time since the Paleozoic.

As mentioned, during the generation of the N-N-E-trending system, magmatic rock activities were extremely violent. Moreover, there were not only large granite intrusions but also eruptions of a large number of neutral-intermediate acid volcanic rocks. These rock masses and rock zones are N-N-E-trending in the second uplift zone and north of the third uplift zone in the Daxinganling region. The current data show at least 4 strong intrusions of granites from the Middle Jurassic to the end of the Early Cretaceous and 5–6 volcanic eruptions from the Late Jurassic to the Early Cretaceous. The above magmatic activities indicate that these large N-N-E-trending uplift zones were generated at the end of the Middle Jurassic or in the Late Jurassic.[15]

However, based on the distribution of deposition strata in certain depression zones and basins, some researchers have argued that the system was generated earlier and probably in the Early Jurassic and initially from the west. Therefore, the generation period of the N-N-E-trending system should be further studied and identified.

Since 1965, five major earthquakes have occurred in the second fold zone of the N-N-E-trending tectonic system in East China. Some investigations have proven that these earthquakes were caused by the continuous torsion of the New Cathaysia fault tectonics. Therefore, some sections of the large N-N-E-trending tectonic system are believed to still be active. The study of the laws and features of earthquakes occurring in this system in East China in recent years may be very important for earthquake prediction in East China in the future.

5.2 New Zealand–Tonga N-N-E-Trending Tectonic System

The system is located in New Zealand and the eastern section of the Pacific Ocean and is composed of a series of N-N-E-trending Mesozoic and Cenozoic basins, including, from west to east, the Coast Basin, East Coast Basin, Canterbury Basin, Soland Basin, New Hebrides Basin, Tonga Basin, etc.

5.3 N-N-E-Trending Tectonic System in the Eastern United States

This system is located in the eastern United States, dominated by the N-N-E-trending Abrachi Mountains, and controls the Abrachi Basin, which is dominated by the Upper Paleozoic and Mesozoic–Cenozoic systems. The sea area is dominated by the Cenozoic system.

This tectonic system is composed of the St. Lawrence Basin, the Xieshelu Basin, the East Coast Basin, the Appalachian Basin and the orogenic zones among them. The southern section is intersected by the northern fault of the Mexico Gulf Basin.

5.4 N-N-E-Trending Tectonic System on the Eastern Coast of South America

This system is located on the eastern coast and offshore South America and is composed of a series of N-N-E-trending Mesozoic and Cenozoic basins. It is dominated by Cenozoic deposits in the offshore region and

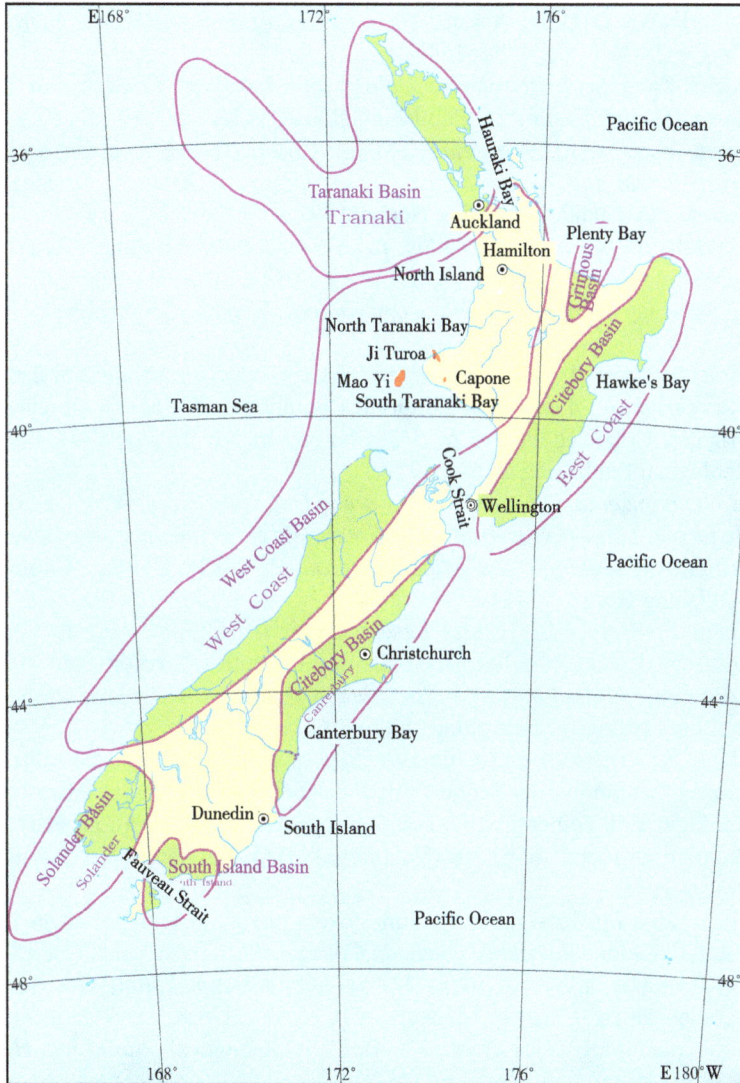

Fig. 5.4.   Schematic diagram of the New Zealand N-N-E-trending tectonic system.

accompanied by a series of Mesozoic and Cenozoic fault zones and magmatic rocks, including the Camps Basin, the Santos Basin, the Pilotas Basin and the Parana Basin (Fig. 5.4).

References

1. Azad, J., Bhattacharyya, S., Datta, B. D. *et al.* Hydrocarbon Accumulation in Nahorkatiya Oilfield, Assam. *Proceedings of the 8th World Petroleum Congress*, 1971, 8(2): 259–268.
2. Yuzhu Kang *et al. Petroleum Geology and Petroleum Resources in Tarim Basin.* Beijing: Geological Publishing House, 1996.
3. Yuzhu Kang *et al. Petroleum Geology Characteristics and Oil and Gas Prospects in the Northwest of China.* Urumchi: Xinjiang Science and Technology Health Publishing House, 1997.
4. Li Dongxu *et al. Introduction to Geomechanics.* Beijing: Geological Publishing House, 1986.
5. Yuzhu Kang *et al. Petroleum Geomechanics.* Beijing: Geomechanics Press, 2013.
6. Li Songshan *et al. New Knowledge of the Arc Top of Huaiyangshan Epsilon-type Structure.* Periodical Editorial Committee of 562 Comprehensive Brigade. Collected Papers of 562 Comprehensive Brigade (#9). Beijing: Geological Publishing House, 1979.
7. Li Zhenxing *et al. Division of Geotectonic Units and Evolution of Geological History in Three Rivers Area in the Southwest of China. Journal of Chengdu Institute of Geology and Mineral Resources* (#13). Beijing: Geological Publishing House, 1991.
8. Liang Jitao *et al.* A Textual Research on the "Chinese Ancient Lands". Editorial Board of the Journal of Nanjing Institute of Geology and Mineral Resources. *Journal of Nanjing Institute of Geology and Mineral Resources.* Beijing: Geological Publishing House, 1991.
9. Liang Jue. *Division and Characteristics of Structural System in Guangxi Zhuang Autonomous Region.* Institute of Geomechanics, Ministry of Geology and Mineral Resources. Collection of Studies on Provincial Tectonic Systems in China (Volume 2). Beijing: Geological Publishing House, 1985.
10. Liu Bo. On Epsilon-type Structural System along the southern edges in the South of China. *Journal of Chengdu College of Geology*, 1982, (2): 45–53.
11. Liu Deliang. *On the Ji-Lu Meridional Structural Belt.* Institute of Geomechanics, Chinese Academy of Geological Sciences. Collection of Geological Mechanics (Part IX). Beijing: Geological Publishing House, 1989.

12. Bakry, G. Approaches to Predicting Reservoir Facies in Frontier Areas — The Alam El Bueib Formation Case Study Obaiyed Area, Western Desert, Egypt. *Proceedings of the 14th Petroleum Conference*, Cairo, Egypt, 1998: 68–83.
13. Balkwill, H. R., Rodrique, G., Paredes, F. I. *et al.* Northern Part of Oriente Basin, Ecuador: Reflection Seismic Expression of Structures, in Petroleum Basins of South America. *American Association of Petroleum Geologists Memoir*, 1995, 62: 559–571.
14. Braidc, S. P. Clay Sedimentation Facies: A Niger Delta Example. *Bulletin of Nigerian Association of Petroleum Explorationists*, 1993, 8(1): 61–72.
15. Brice, S., Kelts, K. R., Arthur, M. A. Lower Cretaceous Lacustrine Source Beds from Early Rifting Phases of South Atlantic. *AAPG Bulletin*, 1980, 64: 680–681.

Chapter 6

N-W-Trending Tectonic Systems

Yuzhu Kang, Shuwen Xing, Zongxiu Wang, Zhijiang Kang, and Huijun Li

Abstract

In this chapter, N-W-trending tectonic systems are introduced, including those in China, central Asia, the North Caucasus, the Zagros of the middle East, the western coast of North America and the Gulf of Suez.

Keywords: Tectonic system; N-W-trending; Type

As early as 1939, Mr. Li Siguang referred to the N-W-trending and N-W-W-trending tectonic systems in western China as the "Western System" to designate the N-W-trending tectonic system in China.

The N-W-trending tectonic system proposed in this book includes N-W-trending and N-W-W-trending tectonic zones with strict regional distributions.

The system is composed of a series of parallel and roughly equidistant N-W-trending compound tectonic zones. The southeastern end of the east Junggar complex tectonic zone is inserted into the Tianshan E-W-trending complex tectonic zone and is compounded obliquely to the Bogda–Haerlik fold tectonic zone of the Tianshan E-W-trending tectonic zone. After the Mesozoic and Cenozoic, due to the combination of the two effects, they were simultaneously involved in the Barkol–Izao tectonic zone and then transformed into a component of the reverse S-shaped tectonic zone.

6.1 N-W-Trending Tectonic System in China

6.1.1 *N-W-Trending Tectonic System in the Junggar Region*

The tectonic system is composed of the Irtysh tectonic zone, the Qiawukar tectonic zone, the Wulungu River–Santang Lack depression zone, the Beita Mountain tectonic zone, the Naomao Lack depression zone and the Krameri–Mochinwula tectonic zone.

(1) Irtysh Tectonic Zone
Southwest of the Altai arc-shaped tectonic zone, a N-W-trending strong compressional tectonic zone is located along the Ertys River. Its main body is confined between the Kizgar–Main'bo Great Fault and the Irtysh Great Fault, with a width of 2,050 km and a length of several thousand kilometers. It enters Kazakhstan to the northwest, extends to Mongolia to the southeast, and gradually widens to the east and west. It presents as a N-W-W-trending low and gentle valley, covered mostly by loose Cenozoic deposits. There are sporadic Paleozoic rocks and blocks exposed between Burqin and the Haba River and Late Paleozoic strata east of Xibodu. The most exposed rocks in this zone are the Late Carboniferous intermediate-acid volcanic rocks and their clastic rocks. In addition, Permian terrestrial coarse clastic rocks are locally exposed, there are very small amounts of Early-Middle Jurassic coal-bearing deposits in the lower reaches of the Haratong Ravine and Qinggeri Rivers, and the Paleogene–Neogene glutenites are distributed on a large scale.[1] There are obvious angular unconformities between the upper Carboniferous and Permian–upper Carboniferous and lower Jurassic systems and between the lower Jurassic and Paleogene–Neogene systems.

The magmatic activities were most intense in the Late Paleozoic (that is, the Late Hercynian) and are dominated by intermediate-acid intrusive rocks. In addition, the basic and ultrabasic rock masses and rock strains are distributed in a zonal pattern, and there are also small Indosinian–Yanshanian acid-basic rock masses. The distribution of the rock masses is approximately consistent with the extension direction of the tectonic zone.

In the Irtysh compound syncline, the overall distribution is basically consistent with the tectonic direction, and the fold axis is undulated and changed. In particular, the axes of small folds are extremely complex, with very different scales, and most of them form skirt-like fold groups. The two wings of the compound syncline are gentle to the north with an inclination angle of approximately 60° and steep to the south with an inclination angle greater than 70°. The granites are invaded along the axis of the folds, and several groups of fault surfaces are invaded in mesh or vein shapes.

The faults in the compression zone are well developed, and strike compression and torsion faults dominate. The N-E-trending and N-W-trending tracing tension fault planes and the near E-W-trending compression and torsion faults are well developed.

The strike-compression and torsion faults are mainly large faults that form the boundary of the compression zone, namely, the Kizgar–Ma'ingbo fault and the Ertys fault. They extend at approximately 290° north to west, showing a gentle wave-like curve, and their cross sections are N-N-E-trending at an inclination angle of approximately 70°. The Kizgar fault is located south of the Altai Mountains, with the middle Devonian system on the northern side of the fault and the upper Carboniferous system on the southern side, at an intersection angle of approximately 30°. There is a broad depression along the fault, and the spring outflow points are distributed in a linear shape. The rock masses on both sides are broken. The fault is 10–100 m wide, and it is filled with a large amount of quartz and calcite clumps, as well as tectonic breccias and fault gouges locally. Some vertical, horizontal and oblique scratches are visible on the fault plane in different directions, indicating that there are many activities and that its properties have changed many times. Both aerial and satellite photographs show clear linear tectonics. There are obvious linear gradient zones on the aeromagnetic anomaly map, and the anomaly axis on both sides intersects at an angle of approximately 30°. The Kizgar fault is staggered to the east by the N-N-W-trending Keketuo Sea fault, with a staggering distance of 25 km to the right. After becoming staggered, the eastern section of the Keketuo Sea fault is called the Mainebo fault. The Irtysh River fault extends along the Irtysh River in the N-W direction.[2] The fault zone is strongly compressed and broken, with a breaking width of 10–50 mm and silicification, sericitization and epidoteification occurring in the zone, showing brown or gray–white strips. Light-colored granite, granite porphyry and quartz veins are widely distributed along the fault zone and mostly fill the secondary fissures in the fault zone in a net or geese shape. The fault intersects the Mesozoic and Cenozoic strata, forming tectonic ridges in the topography. Along the fault, the degree of rock metamorphism is strengthened, and mixed rocks, amphiboles and schists are distributed on the northern side of the fault. On the aeromagnetic anomaly map, there is a N-W-trending positive and negative zone-shaped anomaly on the northern side of the fault and a nearly E-W-trending zone-shaped anomaly on the southern side, with an intersection angle of 15°–20° between the two anomaly axes. Along the fault zone, the magnetic anomaly extends in a clear linear form, and the AT curves are nearly symmetrical. The northern side is slightly gentle, and the southern side is steep. Based on aeromagnetic quantitative calculations, the occurrence of the fault surface is consistent with that in field observations and extends at approximately 75° to the east.

The Irtysh compression zone has a long-term history of activities. It was formed at least in the Early Paleozoic and has important control over the development of the Mongolian and the Junggar arc tectonics, but it matured in the Late Paleozoic. It strictly controlled the distribution of Late Carboniferous strata and caused their strong compression deformation and dynamic metamorphism. The compression zone provided a passage and filling space for the middle-acidic magma activities, especially the intrusions of basic-ultrabasic rocks. The zone was still strongly active in the Mesozoic and Cenozoic.

(2) Qiawukar Tectonic Zone

The zone spreads south of the Irtysh compression tectonic zone and north of the Ulungu River valley. It is composed mainly of Devonian strata, with Ordovician and Carboniferous strata exposed locally. Its western section is connected and compounded with the eastern wing of the Junggar arc tectonic west of the N-N-W-trending Keketuo Sea fault zone, and it is difficult to distinguish the components. East of the Keketuo Sea fault zone, it gradually separates from the Junggar arc tectonic, forming a series of N-W-W-trending compound folds and compression-torsion faults in the Qinggerihe and Burgenhe regions. The Qingri–Burgen anticline, Jiele Karatawu syncline and Chaberti anticline are separated by the N-W-W-trending Karashengar–Jiele Karatawu fault and Kiz–Karaga Ibastau fault. The N-W-W-trending tectonic zone composed of N-W-W-trending folds and compressional faults extends into Mongolia to the southeast and is connected to the N-W-W-trending tectonics in the

Suhaitushan region north of the Hanshuiquan Valley, and they belong to the same tectonic zone.[3]

The middle Devonian system is composed of a set of tuff sandstone, tuff power siltstone and neutral tuff intercalated with biological limestone exposed in the core of the compound anticline. The upper Devonian system is composed of a set of tuff breccia, phyllite, and tuff sandstone, forming the core of a compound syncline. These compound folds are generally symmetrical, with the two wings at an inclination angle of 50°–70°. They are close to the Irtysh compression tectonic zone and the Wulungu River depression from the northern and southern sides, with strong compression, complex tectonics and local syncline folds.

Late Paleozoic intermediate-acid intrusive rock masses are widely developed, most of which extend north and west, consistent with the tectonic line directions. Only a few rock masses are regularly circular.

A group of N-W-trending compression-torsion faults separating these compound anticlines and compound synclines have wide compressional fault zones, with a maximum width of 1 km. The rock masses are obviously faded and ocherized, and there are obvious tectonic steps and valleys. There are more acidic rock veins intruded along the fault, and the rock mass is destroyed by later tectonic activities. These fault zones are favorable zones for the migration and enrichment of copper minerals in medium-temperature hydrothermal fluids.

(3) Ulungu River–Santang Lake Depression Zone

There is a Mesozoic and Cenozoic N-W-trending depression zone from the Ulungu River Valley to the Santanghu Basin in the southeast. The Santanghu region in the southern section of the zone is well developed. Due to the interference of the Almantai Mountain arc-shaped tectonics on the eastern wing of the Junggar arc-shaped tectonics, the morphology of the northern section is more obscure. The Santanghu region in the south has a maximum width of approximately 50 km and a length of 200 km, and the Ulunzhan valley in the north has a width of only a few thousand meters and a length of nearly 100 km.

This depression tectonic zone formed gradually after the Permian on the basis of a large-scale compound syncline composed of Paleozoic strata. The exposed strata in this zone are from the Devonian, Carboniferous, Permian, Jurassic, Cretaceous, Paleogene, Neogene and Quaternary systems. The Devonian system is a multicycle volcanic eruption-deposit formation dominated by intermediate-basic volcanic eruptions that gradually transitioned from the marine facies in the Early Devonian to the marine-terrestrial facies in the middle Devonian and then to the land facies in the late period of the Late Devonian. The marine clastic rocks of the Heitou Formation in the early period of the Early Carboniferous became integrated and covered the Late Devonian strata.[4] An important unconformity occurred between the Dune and Constitutional ages, between the Early Carboniferous and the middle and Late Carboniferous, and between the Carboniferous and Permian.

In the tectonic movements at the end of the Devonian, some uplift tectonic zones of the N-W-trending system underwent strong folding and were basically shaped.

In this depression zone, a series of short-axis and box-shaped folds are present in the Mesozoic and Cenozoic strata dominated by the Jurassic–Neogene system, some domes exist in some regions, and the strata have an inclination angle of less than 15°.

(4) Beita Mountain Tectonic Zone

The fault tectonic zone starts from the Takirbastau on the northeastern margin of the Junggar Basin on the west, passes through the Beita Mountain and the section between the Santang Lake and Nao Lake basins on the east, and is inserted into the Tianshan Mountains E-W-trending compound tectonic zone in the overall strike at 300°–310°. The northern section of the zone is compounded with the Almantai fold zone of the Junggar arc tectonic at an angle of 10°–20°, and some tectonic traces overlap with each other and are difficult to distinguish.

The zone is composed mainly of Silurian, Devonian and Carboniferous strata. There is a very small exposure of the Silurian system at the core of the Kupu compound anticline, which comprises a set of shallow metamorphic marine clastic rocks with a small amount of basic extruded rocks. The Devonian system is the most widely distributed stratum and constitutes the main body of the zone. The lower Devonian and Silurian systems are integrated together, but the upper section of the lower Devonian system is overlaid on the Early Paleozoic granite masses. The tectonic movements between the early and late Paleozoic were delayed here. The upper, middle, and lower Devonian systems are well developed in this zone. However, they show two different lithofacies. One is represented by the Beita Mountain region, in which there are mostly neritic volcaniclastic and volcaniclastic deposits, as well as some volcanic lava locally. The other is represented by the Renge Iridon region north of Takirbastau and is dominated by normal clastic rocks. The upper, middle and lower Devonian systems transformed gradually from sea-shed facies to sea–land interaction facies and then to land facies. The two lithofacies regions are bounded by a large N-W-trending fault, reflecting that the N-W-trending system was strongly active in the Devonian and strictly controlled the distribution of volcanic rock zones and normal deposit zones in this zone. By the Carboniferous, there was no obvious differentiation, and they the rocks all volcanic clastic rocks composed of intermediate-basic volcanic rocks formed in multiple eruptive cycles. After the Early Carboniferous, the history of transgression basically ended, and a N-W-trending compound anticline formed.

In general, the Beita Mountain fault tectonic zone is composed of a series of compound anticlines with a length of 400 km. From west to east, it includes the Derenge Yirideng compound anticline, the Hassafen compound anticline, and the Kupu–Beita Mountain compound anticline, with a fold axis at 290°–310°. The core of the compound anticlines is composed of Silurian and Early Devonian strata, and the two wings are composed of middle and Late Devonian and Carboniferous strata. The folds are tightly closed and symmetrical, with some local reverse folds, and the inclination angle of the two wings is greater than 70°. There are slight geese-shaped arrangements between the compound anticlines, and there are large intermediate-acid rocks in the cores of the compound anticlines.

There are developed fault tectonics in the Beita Mountain fault tectonic zone, dominated by large-scale N-W-trending compression-torsion faults. These main fault planes are generally inclined northeast and are associated with one set of torsion fault planes at 320°–330° north to west and another set of torsion fault planes at approximately 50° north to east. They are generally not large in scale and are often densely distributed on the two wings of the anticline and intersect each other. The most important controlling faults in this zone include the Crovin–Shanghu fault, the Kupu–South Beita Mountain fault and the Takirbastau fault.

Crovin–Shanghu fault: This fault starts from the northern side of the Derenge Yirideng compound anticline on the northwest and passes through the Beita Mountain region to Sanyong Lake on the southeast, covered by Quaternary loose deposits. The overall trend of the fault zone is a gentle wave-like curvature. It is often an extrusion zone composed of several dense parallel faults, with a maximum width of 4 km and a typical width of 2–3 km. The rocks in the zone are strongly schistosified and phyllized, with local mylonite zones. The main cleavage surface and schistose surfaces are parallel to the direction of the extrusion zone. There are a large number of quartz veins and calcite veins in the zone, as well as a large number of intermediate-acid rock branches and dykes locally. The fault appears as a negative topography distributed in a line, with obvious distinguishing features in aerial and satellite photos. The fault is generally N-E-trending, with an inclination angle greater than 80°, and some sections are S-W-trending, with a total length of 400 km.

Kupu–South Beita Mountain fault: This fault spreads at 290°–305° north to west and shows a gentle wavy curve. It is N-E-trending, with an inclination angle of 60°, a width of over 500 m, and a long axis direction of compressed flakes and cleavage. The tectonic lens in the fault zone is parallel to the fault zone, with obvious images on aerial photos and satellite photos. The entire fault zone is exposed intermittently, and its middle and southern sections are covered by Quaternary deposits.

Takelbastao fault: This fault is located on the northeastern Junggar Basin and on the northern margin of the Takeer depression. It is the piedmont fault of Takelbastao. It spreads at 300° and is inclined to N-E-trending, with an inclination angle greater than 60°. The compressed fault zone has a width of 100–350 m and is filled by a small amount of quartz veins, calcite veins, and diorite porphyry veins parallel to the fault direction. The rocks are brown and ocherized, and springs are exposed along the fault zone.

The axes of the two groups of N-E-trending and N-W-trending conjugate torsional faults are curved in wavy form along the entire fault tectonic zone, and the directions are slightly changed in different sections. The former changes from N-E-trending to N-E-E-trending, while the latter changes from N-W-trending to N-N-W-trending. They are generally not large in scale, with lengths of approximately 30–50 km. The N-E-trending and N-E-E-trending faults are torsional and compressible, while the N-W-trending and N-N-W-trending faults are torsional and tensional. They all obliquely intersect the direction of the tectonic line, with significant horizontal displacements. The former shows left-hand torsion, and the latter exhibits right-hand torsion, sometimes with a breakage distance of 5–10 km. These two groups of conjugate torsion faults are best developed in the two wings of the Beita Mountain compound anticline.

(5) Naomao Lake Depression Zone

This zone is a discontinuous negative tectonic zone sandwiched between the Beita Mountain fault tectonic zone and the Moqinwula–Kelamaili fault tectonic zone. It is gradually new from northwest to southeast; that is, it is composed of Triassic–Jurassic–Cenozoic basins in a successive development sequence. The southern section is dominated by the Naomao Lake Basin, passes through the Beita Mountain salt pond region to the northwest and passes to the Takeer depression northeast of the Junggar Basin.[5] The two Takeer depressions form a large pocket that is open to the northwest, located between Klamali Mountain and Takir Bastau Mountain. In the middle of the Junggar Basin, this N-W-trending depression zone was discovered through geophysical data. This depression was deposited before the Triassic. In the southeast, a good coal-forming basin formed in the Jurassic in the Beita Mountain Salt Lake Basin. In the Naomao Lake region further southeast, it is mainly a N-W-trending basin formed after the Paleogene, with shallow coverage. The Paleogene, Neogene and Quaternary glutenites are directly unconformable on the Carboniferous strata, and the Jurassic and Cretaceous systems are only distributed on the northern margin and in the depression of the Naomao Lake Basin. In the Takeer Depression and the Naomao Lake Basin, the Paleogene–Neogene strata weakly folded, forming a series of N-W-trending short-axis anticlines, N-W-trending nose tectonics and dome tectonics. In the axial of the anticlines or dome tectonics, the strata extend obliquely at 20°–30°, gradually slow down, and even tend to be horizontal toward both wings.

(6) Krameri–Moqinwula Tectonic Zone

The tectonic zone is dominated by the N-W-trending the Moqinwula Mountain. It extends to the Karamai Mountain in the northwest and then enters the Junggar Basin. The basin's internal basement and the depth map from geophysical data clearly show that this N-W-W-trending uplift tectonic zone extends obliquely through the central Junggar Basin and then reaches the margin of the West Junggar Mountains, which is conversely compounded with the western wing of the Junggar arc tectonics. The southeastern section of this tectonic zone is inserted into the Tianshan Mountain E-W-trending compound tectonic zone and is compounded obliquely with the Awulali–Bogda–Haerlik fold tectonic zone. It was involved in the Barkun–Iwu reverse S-shaped tectonics after the Mesozoic and Cenozoic, and most components were transformed into components of the reverse S-shaped tectonics.

The tectonic zone is composed mainly of Ordovician, Silurian, Devonian and Carboniferous systems. The Ordovician and Silurian systems are a set of shallow metamorphic carbonate rocks, intermediate-acid volcanic rocks and pyroclastic rocks.[6] During the Devonian, the deposit facies and the geographical environment were significantly differentiated due to the separation of the N-W-W-trending deep faults. The crustal depression was very violent north of the N-W-W-trending deep fault, accompanied by large-scale seafloor fissure eruptions. The rock natures change drastically from volcanic facies to volcanic clastic rocks and volcanic deposit clastic rocks along the strike. There are relatively stable normal deposit rocks south of the Krameri deep fault, and there is only a small amount of volcanic activity locally, with a low deposition rate, small deposition thickness and relatively stable lithofacies. This kind of strong activity and relatively stable zone differentiation caused by the separation of N-W-W-trending deep faults are obviously related to the strong activities of the two systems. After the Early Carboniferous, this kind of differentiation was no longer obvious, and the rocks became a set of volcanic clastic rocks and normally deposited clastic rocks, accompanied by strong intermediate-acid volcanic activity in some regions. After the late period of the Early Carboniferous, most regions were uplifted. Except for some continuous deposits in some regions on the southern section, other regions were corroded in the long term and showed the basic shape of the N-W-W-trending fold fault tectonic zone.

The northwestern section of the fault tectonic zone comprises the Karamayli folds, and the southeastern section comprises the Moqinwula compound anticline. The Krameri folds are composed of a series of N-W-trending anticlines and synclines, with lengths of 20–40 km and widths of 4–5 km, and they are composed of Devonian and Carboniferous strata. The secondary minor folds are very complex, with an inclination angle of 60°–80° on both sides. They are partially affected by the fault, and the axial planes are reversed to the south, forming synclinic folds dipping to the north. The core of the Moqinwula compound anticline is composed of Ordovician strata, and the two wings are composed of Devonian and Carboniferous strata. A series of N-W-trending large faults cut the tectonic zone into many imbricate-like fault blocks, presented as monoclinic (syncline) zones, but on the whole, the appearance of an anticline is still preserved. It is just that the southern wing was cut more incompletely. All rock formations are tightly closed linear folds in the N-W-trending and S-E-trending directions, with an inclination angle of 40°–60°.

The secondary folds are very complex, forming S-W-trending synclines and N-E-trending reverse folds.

The large-scale compression or compression-torsion faults in this zone include the Karameili fault and the Moqinwula fault. All these faults are N-W-W-trending and S-E-E-trending, with lengths of hundreds of kilometers. Among them, the Shiying Beach–Kelamaili Mountain North fault is connected to the Northeast Barkun Basin fault on the southeast and extends through the center of the Junggar Basin on the northwest and to the southeastern margin of the West Junggar Mountains, with a total length of 600 km.[7] Most of the faults are highly reversed faults, with an inclination angle greater than 60°, and some faults are partially overridden. Due to the steep occurrence of the fault planes, the trend of the fault planes is often unstable along the strike from northeast to southwest, and this group of faults all have wider fault zones, with obvious images in aerial, satellite and geophysical anomaly maps. This group of N-W-trending faults all have a long history of activity and a large cutting depth in the crust. Most of them are composed of basic and ultrabasic rocks and control the Paleozoic volcanic eruptions, the intrusions of a large amount of intermediate-acid magma, and the formation of gold-bearing quartz veins in the Krameri region.

In this tectonic zone, two sets of torsion fault planes are matched with the N-W-W-trending compressional tectonic planes, namely, the NW-NNW-trending fault group and the NE-NEE-trending fault group. They have a small scale, a length of 5–20 km, an oblique stratigraphic trend and vertical fault planes. The former is left-hand torsion, while the latter is right-hand torsion, with a breakage distance of 100–200 m.

In the Kelamaili region in the north, because it is compounded obliquely with the Liukeshu–Laoyamiao E-W-trending tectonic zone, the E-W-trending folds and faults are also relatively developed, but the scale is small. The E-W-trending fault zones are more active, often intersecting several groups of faults in other directions.

6.1.2 *Boroholo N-W-Trending Tectonic System*

This tectonic system is a large-scale N-W-trending complex tectonic system that is obliquely connected to the Tianshan Mountain E-W-trending compound tectonics, dominated by the Boroholo–Kavabulak fold fault tectonic zone, accompanied by the Aiding Lake depression tectonic zone to the north and the Yili-Yanqi depression tectonic zone to the south.

The underdeveloped Saalming–Koketieke fold fault tectonic zone has a length of 1,100 km from east to west. They are distinguished from the E-W-trending tectonic system in their unique and regular combination of tectonic traces, stable tectonic line direction, large scale and unique tectonic development history to make the geological tectonics of the Tianshan region more complicated. Pro. Li Siguang pointed out many times that the Tianshan Mountain E-W-trending tectonic zone moved northward due to interference from other tectonic factors. Therefore, the combination of the N-W-trending system and the E-W-trending tectonic system in the Tianshan region is the main factor in the northward movement of the western section of the Tianshan Mountains.

The Boroholo compound tectonic zone can be divided into five secondary tectonic zones from north to south.

(1) Aibi Lake–Aiding Lake Depression Zone

This zone spreads at the northern foot of the Tianshan Mountains and on the northern side of the Boroholo–Kavabulak fold fault tectonic zone. It is compounded obliquely with the two depression tectonic zones of the E-W-trending tectonic system (namely, the Wusu–Qitai depression tectonic zone and the Yili–Hami depression tectonic zone) on the southern margin of the Junggar Basin and the western side of the Turpan Basin. The negative composite sections of the two major tectonic systems have the deepest depression, which are the two large lake depressions. The composition of the northern section of the depression zone and the Wusu–Qitai E-W-trending depression zone forms Aibi Lake, and its composition and the Yili–Hami depression zone form the lowest inland lake in China — Aiding Lake (with an altitude of 154 m).

The depression tectonic zone generally extends in the N-W-W-trending direction, but the basins and depression centers in the zone extend in the E-W-trending direction. From northwest to southeast, they are the Aibi Lake depression, Manasi depression, Chaiwopu Lake depression and Aiding Lake depression, with a spacing of 150–200 km. They are arranged obliquely from northwest to southeast, forming a ξ-shaped depression zone.[8]

This depression tectonic zone has gradually developed since the Permian and is dominated by Permian and Mesozoic–Cenozoic strata. The lower Permian system is a set of littoral facies and sea–land interaction gray–green clastic rocks, unconformably covering the Carboniferous system at a significant angle. The Upper Permian and Lower Triassic systems are composed of red clastic rocks dominated by purple–red and gray–green conglomerate, sandstone, and mudstone. The Middle–Upper Triassic system is composed mainly of gray–green mudstone and coal-containing sandstone. The lower Jurassic system is composed of gray–green sandstone, mudstone, conglomerate and coal seams, which are mainly coal-bearing formations. The upper Jurassic system is composed of red mudstone, sandstone, and conglomerate. The lower Cretaceous system is dominated by lacustrine variegated strips of argillaceous rocks, while the upper Cretaceous system is composed of fluvial clastic rocks. The Paleogene–Neogene system is composed of red clastic rocks.

(2) Boroholo–Kavabulak Fold-Fault Tectonic Zone

The zone starts from Boroholo Mountain and Keguqin Mountain on the northern margin of the Yili Basin in the west, passes through Yilianhabirga Mountain to the Kavalbraktag region in the east and is greatly weakened due to the limitation of the strong original E-W-trending tectonic zone (namely, the Lateke–Kuruktag–Alatage fold fault tectonic zone) on the eastmost. At the same time, it is intersected by the N-E-trending tectonic zone of the Altyn system. In the Beishan region, only some fragments are contained in the N-E-trending tectonic zone of the Altyn system and correspond to the Qilianshan N-W-trending tectonic zone across the Hexi Corridor in Gansu Province.

This zone has a length of 1,300 km in Xinjiang, and its main body is composed of the Boroholo compound anticline, the Yilianhabirga compound anticline, the Kavabulak compound anticline and the Kiziltag compound syncline.

(a) *Boroholo compound anticline*: The Boroholo compound anticline starts from the shore of Sailimu Lake in the west and extends to east of Shengli Daban, showing a N-W-trending long and narrow strip. The core of the compound anticline is composed mainly of the Sinian subsphere and the Paleoproterozoic systems, and the Silurian strata are most exposed. The strata generally have different degrees of metamorphism, dominated by moderate-shallow metamorphic marine carbonates and clastic rocks, forming N-W-W-trending or nearly E-W-trending closed linear folds. Most of the folds are asymmetrical and sometimes reverse in the southwestern direction. The two wings of the fold have an inclination angle of approximately 40°–80°. The low-level small folds in the core and the wings of the compound anticline are very complicated. The two wings are composed of Devonian–Carboniferous shallow metamorphic pyro-clastic rocks, volcanic lava and carbonate rocks.[9] The axis of the fold and the two wings are intersected by a series of large N-W-trending compression-torsion faults, showing discontinuities and incompleteness.

In the Keguqin Mountain region on the northwestern end of the compound anticline, the core is composed of a series of Precambrian and Lower Paleozoic nearly E-W-trending anticlines and synclines and some nearly E-W-trending compressional faults arranged obliquely in ξ-shaped form from northwest to southeast (Fig. 6.1).

Each secondary fold has a length of 50–80 km, and the fold axis is often turned reverse to the south. The fold axis plane and the compressive fault surface are inclined to the north at an inclination angle of 50°–70°, with an interval of approximately 20 km between the two anticline axes. This ξ-shaped tectonics clearly reflect the right-handed torsion characteristic of the N-W-trending tectonic zone. At the same time, the E-W-trending folds in the core of the Keguqin Mountain compound anticline composed of Precambrian strata may reflect the features of the early E-W-trending tectonics and constitute a typical example of the

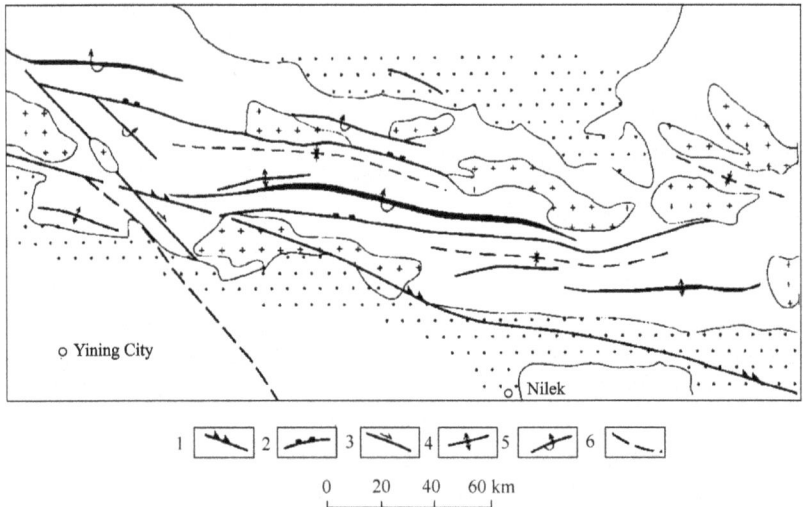

Fig. 6.1.   Keguqin N-W-trending ξ-shaped synclines.

Notes: (1) Compressive and torsional large fault; (2) Compressive fault; (3) Torsional fault; (4) Anticline; (5) Compound reverse anticline; (6) Satellite-interpreted fault.

N-W-W-trending compound syncline superimposed on the E-W-trending tectonics. The Devonian–Carboniferous N-W-W-trending strata on the two wings of the compound anticline are obviously unconformably covered, with a large intersection angle between them.

In the southeastern section of the Boroholo compound anticline, the exposed axis is narrower. The ξ-shaped arrangement is not obvious, dominated by N-W-W-trending closed linear folds, reflecting stronger NE-SW-trending compression and relatively weak right-handed torsion.

(b) *Yilianhabirga compound anticline*: The core of the compound anticline is composed of the middle Devonian system, and the two wings are composed of Carboniferous strata, forming a complete compound anticline tectonic. However, due to being intersected by a series of large compression and torsion faults with the same trend as that of the fold axis, the tectonics have been destroyed. In general, they should form a large secondary fold on the northern wing of the Boroholo compound anticline.

Yilianhabirga compound anticline: The middle Devonian system in the core is dominated by a set of phyllitized gray–green-gray fine volcanic clastics with a thickness of nearly 10,000 m and a small amount of clastic rock and carbonate rock deposits. The southern wing of the compound anticline is basically broken and damaged and is composed of only lower Carboniferous gray–black clastic rocks, limestones and tuffs retained in the form of long and narrow fault blocks. The northern wing of the compound anticline is well preserved and is dominated by the middle Carboniferous system composed of a set of gray–black volcanic clastic rocks, and the lower Permian system is composed of a set of purple–red to gray–green mid-basic and mid-acid volcanic rocks. The internal secondary folds of the compound anticline are complex and are strongly compressed, with an inclination angle of 60°–85°. The large compression-torsion faults often extend N-W-W-trending in line with the strata strike, forming a N-S-trending shingled fault system on the northern wing of the compound anticline. In the Yilianhabirga compound anticline zone, intrusive rocks are not developed, and there are no large intrusive rocks exposed; rather, only vein-like, long-striped or string-like ultrabasic rocks are distributed near the large compression torsion fault zone, forming a better jade ore zone.

(c) *Cavabrac compound anticline*: The compound anticline is composed mainly of the Sinian subboundary rock system. The upper section is composed of a set of carbonate rocks, called the Kavabulak Group, and the lower section is composed of a set of gneiss and crystalline schist, called the Xingxingxia Group. The northern wing of the compound anticline was intersected and damaged by the Kava Brak fault; the southern wing was composed mainly of the Devonian system with a large thickness. The lower section was composed of clastic rocks, and the upper section was composed of carbonate rocks. The compound anticline is generally N-W-W-trending. In its southeastern section in the Garbulak region, it is intersected by the Palgun fault of the Altyn system, resulting in a relatively large leftward torsion, southward movement of the western section, and northward movement of the eastern section of the anticline. The secondary folds of the compound anticline are relatively complete N-W-W-trending closed folds, with the fold axis inclined westward (with an inclination angle of 30°–40°). The axis plane is inclined to the south. The northern strata are steep, with an inclination angle of 70°–80°, and are often reversed to the south. The southern wing is gentle, with an inclination angle of 40°–60°.

(d) *Kizildag compound syncline*: The syncline is located southwest of the Kavabulak compound anticline and northeast of the Yanqi Basin, including the Wutonggou–Wollonggang compound syncline extending to the southeast.

The Kizildag Syncline is generally N-W-trending, and it turns N-S-trending in the Wollonggang region on the southeastern end due to the interference of shrimp-shaped mountain zonal tectonics. The lower Carboniferous and middle-upper Devonian systems in the core of the compound syncline are composed mainly of carbonate rocks and some clastic rocks. The two wings are composed of lower Devonian and Silurian shallow metamorphic phyllites and schists. The Early Carboniferous strata often unconformably cover the Devonian system at an angle and exist in the core of the syncline. In addition to being controlled by N-W-W-trending tectonics, they are also controlled by the E-W-trending tectonic system.[10] The secondary folds in the compound syncline are more complicated. The single anticlines and synclines are still nearly E-W-trending, but they form a N-W-W-trending ξ-shaped oblique tectonic zone generally from northwest to southeast, reflecting the right-hand compression and torsion nature of the N-W-W-trending tectonic zone.

The paleogeographic analysis of lithofacies shows that the compound syncline developed on the basis of a deposit trough formed after the end of the Early Paleozoic. The Kavabulak compound anticline was compressed and uplifted at the end of the Silurian, and a N-W-W-trending trough was formed on the southern side of the uplift in the Devonian. The Early Devonian system was dominated by terrigenous clastic deposit rocks, shallow sea carbonate formations were deposited in the middle Devonian, and strong volcanic activity occurred in the Kiziltag region in the Late Devonian, depositing a set of intermediate-acid volcanic rocks and pyroclastic rocks. However, there are still terrigenous clastic rocks and shallow sea carbonate rocks in the southeastern Hazierbulak region. The Devonian deposits have a thickness of over 5,000 m. Such a long and stable semien closed narrow trough in a depression state has created a good paleogeographic environment for the formation of iron ore deposits and has become an important metallogenic zone controlled by the N-W-trending system.

In addition to the abovementioned compound anticlines and synclines, the Boroholo–Kavabulak fold fault tectonic zone has developed a series of large N-W-trending parallel compression and torsional faults, especially for the Wulastai fault (namely, the large fault on the northern margin of the Tianshan Mountains or the large fault on the southern margin of the Urumqi depression), the Yamate fault, the Kekebastao fault, the Lante Daban fault, the Shengli Daban fault (namely, the Yilianhabirgada fault), the Pine Tree Daban fault (namely, the Boroholo fault), the Kokqin fault, the Maanqiao fault, the Kavabulak fault, the Wuyongbuke fault, etc., with lengths from several kilometers to hundreds of kilometers. Most of their fault surfaces are inclined toward the core of the compound anticline, pushing outwards layer by layer and forming an imbricated tectonic, especially the northern wing of the Boroholo compound anticline. These large-scale compressional faults are nearly parallel to the folds, and they all have wider broken zones, forming wider dynamic metamorphic zones. The flakes, cleavage, extruded lens and small associated folds in the fault zones and the dynamic metamorphic zones are nearly parallel to the main fault or intersect at an angle below 15°, clearly showing the strong compression or compression torsion feature of the main fault. At the same time, a series of arc-shaped tectonics, torsional tectonics and herringbone tectonics on both sides of the fault zone and ξ-shaped tectonics formed by oblique fold rows uniformly reflect the strong right-hand torsion nature of these large N-W-W-trending faults.

These large N-W-trending faults generally have a long history of activities, controlling the distribution of Silurian and Devonian deposits, intersecting the Paleozoic and pre-Paleozoic strata, partially intersecting the Paleogene–Neogene and Quaternary strata, and controlling various types of intrusive rock masses formed at different ages, and the ultrabasic rocks are generally distributed along the fault zone. The morphology of Mesozoic and Cenozoic deposit basins and the distribution of recent seismic activities and hot springs in the N-W-trending fault zone show that these faults were still very active in the latest period.

In the Boroholo–Kavabulak tectonic zone, multiple types of intrusive rock masses are widely developed in multiple periods and are distributed mainly in the cores of the Boroholo compound anticline and the Kavabulak compound anticline.[11] The rock mass is NW-SEE-trending, which is consistent with the direction of the regional tectonic line. The Proterozoic and Early Paleozoic intrusive rocks are mainly granites (γ_2 and γ_3), and they generally exhibit different degrees of metamorphism. The main rocks include gneissic granite, mylonite, granite, plagioclase gneiss, mixed rock, etc., with a small distribution range and a small scale. The extensive and large-scale exposures are a variety of Late Paleozoic intrusive rocks. Several granite isotope samples collected from Kumish to Kavabulak are concentrated mainly in these three periods: 410–439 Ma, 333–350 Ma and 225–248 Ma, corresponding to the Silurian, from the end of the Devonian to the Carboniferous and the end of the Permian, respectively, including ultrabasic to acidic rock masses. The ultrabasic, basic and neutral intrusive rock masses are generally small in scale, in the form of rock strains, rock branches and rock caps, distributed along the fault zone in the form of beads. The intermediate-acid intrusive rock masses appear as mainly large-scale rock foundations, with the distinctive feature that the shape, extension direction and arrangement of the rock masses are strictly under the composite control of the N-W-trending tectonic system and the E-W-trending tectonic system. The compressional fault zone between the Shengli Daban fault and the Songshu Daban fault provides a channel for magmatic rock intrusion and activity space, forming a tectonic magmatic rock zone with a width of 30–50 km and a length of nearly 1,000 km. This tectonic magmatic rock zone is also an important ore-controlling tectonic zone in the Tianshan region. In particular, the combined section with the E-W-trending tectonic system and the Altyn tectonic system forms the most favorable metallogenic section in the Tianshan region.

(3) Qilian Mountain Tectonic Zone

This zone is distributed in the Qilian Mountains and its southern and northern margins. It formed by tectonic movements at the end of the Silurian. The Devonian system was angularly unconformable in the Silurian. A series of folds and thrust faults at 320°–330° north to west formed in these movements, together with rock blocks and fold zones sandwiched between them, forming a N-W-trending tectonic system. From north to south, they are the Longshou Mountain fold fault zone, the Corridor compound syncline zone, the Corridor South Mountain–Lenglongling fold fault zone, the Tuolai South Mountain–Datong Mountain fold fault zone and the Wulandaban Mountain–Laji Mountain compound syncline zone.

The tectonic morphologies of the N-W-trending system in this region are quite different from those of the Qiluepsilon-shaped tectonic system. The system exhibits strong folds, mostly in compact linear form, with considerable length, a large scale, and associated compression and compression and torsional fault groups parallel to the folds. The Early Paleozoic deposits of this tectonic system are dominated by thicker clastic rocks and carbonate rocks sandwiched by a large amount of intermediate-basic volcanic rocks, with a total thickness of 20,000 m, of which the volcanic rocks formed during 436–445 Ma (Xia Linqi *et al.*, 2001).

The Cambrian–Silurian system is intruded by acidic rock masses, the zircon U–Pb age is approximately 450 Ma and its direction is consistent with that of the abovementioned faults and folds. It also controls several Early Paleozoic basic and ultrabasic rock zones in this region and is distributed near the fold axis of the system. Among them, the zircon age of the Shijuli gabbro in the western section of the North Qilian Mountains is 457.9 Ma (Song Zhongbao *et al.*, 2007), and the isotopic age of the basic volcanic rocks in the Baiyin ore field is 465 Ma (Li Xiangmin *et al.*, 2009).

(4) Yining–Yanqi Depression Tectonic Zone

The Yining–Yanqi depression tectonic zone is located on the southern side of the Boroholo–Kavabulak fold fault tectonic zone. It is composed of a series of small-scale Mesozoic and Cenozoic basins, i.e., the Yining Basin, the large and small Yurdos Basins and the Yanqi Basin from west to east. The zone is generally N-W-W-trending, but single basins are nearly E-W-trending. The margins of the basins are controlled by E-W-trending, N-W-trending and N-E-trending faults, and they are often rhomboid and triangular basins, arranged obliquely in a ξ shape from northwest to southeast.[12] These basins are developed on the Paleozoic fold basements. The interior of the basin is composed mainly of Jurassic, Paleogene–Neogene and Quaternary deposits.

The folds and uplifts in the region were eroded by strong crustal movements at the end of the Permian. There were basically no deposits in the zone in the Triassic, and a set of river, lake, and swamp facies of coal-bearing formation deposits settled on the basically flattened foundation in the Tianshan region in the Late Jurassic. Due to the humid and warm climate, these deposited basins have good coal-forming conditions. Because the Tianshan Mountains were uplifted again in the Cretaceous, Cretaceous deposits were generally absent in the zone and basins. The Jurassic basin was basically inherited in the Paleogene–Neogene, but the deposit range was further expanded, forming a substantially connected depression tectonic zone dominated by red and brown sandstone, mudstone and conglomerate masses. From the end of the Neogene to the Early Quaternary, the Tianshan Mountains were raised on a large scale in the Himalayan Movements, elevating both the large and small Yurdos Basins to an altitude of 3,000 m. The connected Paleogene–Neogene basins are divided into a series of smaller basins due to erosion and cutting in the Quaternary, clearly showing a ξ-shaped arrangement, especially in the eastern section of the Yanqi Basin.

Since the Quaternary, this unified Paleogene–Neogene basin has been disintegrated and divided into the Bosten Lake depression, the Wuzongbula depression and the Kizile depression. The Kizile depression is a N-W-trending uplift zone between the two depressions. The interior is composed of a series of nearly E-W-trending lower-level depressions and uplifts. Each of these subbasins has a length of 80–150 km and a width of 10–15 km, and the interval between the two basins is equivalent to the width of the basin.[13] These basins and uplifts are arranged obliquely from northwest to southeast. At least six relatively complete basins form a perfect ξ-shaped tectonic zone, showing that the depression tectonic zones of the N-W-trending system were recently revived in the Neoid and that there was obvious right-hand torsion. The margins of these uplifts and depressions are often controlled by compressional faults. On the southern margin of Kiziltag, the Carboniferous strata were overlaid on the

Paleogene–Neogene strata, indicating that this ξ-shaped tectonic feature formed or that the unified Paleogene–Neogene depression basin was disintegrated after the Paleogene–Neogene.

(5) Saalmin–Koktieke Fold-Fault Tectonic Zone

The zone spreads southwest of the Bosten Lake Depression, north of the Tarim Basin, and in the middle section of the South Tianshan Mountains. It is composed mainly of the Saalmin compound anticline in the north, the Hula Mountain extrusion zone in the middle, and the Keketike compound syncline in the south. There is no clear boundary between its northern section and the Yining–Yanqi depression zone. The Yining–Yanqi depression tectonic zone developed after the Mesozoic and Cenozoic, is superimposed on the Paleozoic tectonic zone, and is a tectonic layer related to the northern section of the Saalmin–Koketieke fold fault tectonic zone in genetic relations of the upper and lower sections but with different tectonic periods.

(a) *Saalmin compound anticline*: The compound anticline is N-W-trending and is composed mainly of a set of middle Devonian carbonate rocks and marine clastic rocks with a thickness of 5,000–8,000 m. The axis is undulated eastward, and the two wings are composed of a series of secondary anticlines, synclines and interlayer folds. The axes of these secondary folds are mostly inclined northward, showing a tightly inclined and reverse closed fold shape. The larger faults in the compound anticline are also N-W-W-trending and are composed mostly of high-angle thrust faults. Intrusive rocks are not developed, and only large granite bases (r_4^1) are exposed northwest of the Bosten Lake Depression, with the long axis consistent with the direction of the main tectonic line.

(b) *Hula Mountain compression zone*: The zone is sandwiched between the N-W-trending Hula Mountain fault and the Kaidu River fault, with a total length of 200 km and a width of 8–10 km. It is narrow in the middle and wide at both ends. There are obvious differences between the eastern and western sections of the compression zone. The eastern section is composed mainly of a set of ancient metamorphic rock systems, i.e., Sinian subboundary Changcheng schist and gneiss, forming a N-W-W-trending compressional metamorphic zone. A variety of intrusive rock masses are present in the compression metamorphic zone and the closed fold axis, schists gneiss, long-strip intrusive rock masses and the main compression faults in the zone are uniformly N-W-W-trending.[14] The western section of the Hula Mountain compression zone is composed mainly of a set of thick upper Carboniferous argillaceous siltstones and a small amount of exposed Permian thick continental strata. The large thick Late Carboniferous deposits may have formed by depressions in the faults on both sides of the compression zone after the early tectonic stress release of the strong compression. However, the Late Carboniferous strata are also strongly compressed, the folds are very complex, and the small folds are very developed. There are common folds in various forms of inversion, sharp margins, and horizontal lying, forming a large asymmetric anticline tectonic, and the anticline axis is inclined to the south.

(c) *Koktek compound syncline*: This compound syncline is composed mainly of middle-upper Carboniferous tuff sandstones and tuff rocks. The middle section of the syncline is N-W-trending, and the two ends turn E-W-trending in a slight reverse S-shape overall. The folds are complex and mostly closed linear ones, and the two wings are basically symmetrical, with a relatively steep shape and an inclination angle generally greater than 70°. Some regions on the southern wing show a trend of inversion to the south to different degrees, and the secondary folds of the syncline are developed and form a high-angle reverse fault. The intrusive rock activities are weak, but the ultrabasic-acid intrusive masses are exposed in the shapes of rock strains and rock branches, strictly controlled tectonically and distributed in regions where faults or fissures are developed, where the distribution of metal minerals is also controlled by these fault magmas.

6.1.3 *Awati–Mangal–Qaidam Depression Zone*

The depression zone is located northeast of the Tarim Basin and passes the Altyn tectonic zone to the southeast, forming a depression zone with the Qaidam Basin. It formed in the Early Paleozoic and continuously subsided from the Late Paleozoic to the Cenozoic. It is an important depression zone in the northwestern region and an important petroleum-bearing region.

6.1.4 *Bachu–Qimantage–Supilin Tectonic System*

This tectonic system is located in the Bachu region in the middle of the Tarim Basin and passes the Altyn Mountains and the Qimantag Formation to the southeast, forming an uplift zone.[15]

(1) Bachu–Khattak Uplift Zone

This zone is located in the middle and north of the Tarim Basin, was formed in the Ordovician, was strongly uplifted in the Late Paleozoic and continued to the Mesozoic. Mesozoic deposits were missing in it. The Paleogene–Neogene system directly covers both sides of the Carboniferous–Permian uplift and is controlled by thrust faults. Its main body is N-W-trending.

(2) Qimantag Fold Tectonic Zone

This zone is located south of Altyn, north of Burhanda Mountain, and between the Kumukuri Basin and Qaidam Basin. The overall tectonic extends at 300° north to west, with a width of approximately 70 km from south to north and a length of approximately 450 km from east to west. Overall, it is a reverse S-shape, with its southeastern section located in Qinghai Province.

This fold tectonic zone is dominated by a fold tectonic composed of thicker Late Paleozoic marine clastic rocks and carbonate rocks and integrated with the tectonic masses of Pre-Sinian and Lower Paleozoic green phyllite, schist interbedded with marble, and volcanic rocks, forming a compound anticline, on which a small number of small Paleogene–Neogene deposit basins are superimposed.

The fold tectonics composed of the Paleozoic and previous strata are mostly linear folds, which are arranged in geese in some regions, and complementary N-W-W-trending fault tectonics are relatively developed. The fault planes are inclined southwest with compression or compression torsional features. Both sides of the fault often exhibit Late Paleozoic acid, basic, and ultrabasic rock intrusions, as well as sporadic distribution of Mesozoic granite rocks. Satellite photographs have shown some derivative torsion tectonics with clearer images on the side of large-scale faults, indicating that torsional activity has occurred in the faults. The long-term inherited activities in fault tectonics not only changed the distribution features of the pre-Mesozoic strata to some extent but also caused geological phenomena in which the Paleozoic strata overwhelmed the

Paleogene–Neogene red beds. In addition, the N-E-trending and N-W-W-trending faults associated with the N-W-trending faults are also well developed. They are genetically cross-linked; the former is not very obvious in torsion activities, and the latter has obvious features of right-hand torsion, with an obvious fault distance of 2–3 km. These two groups of torsional fault tectonics have strict control over the boundaries of modern topography and the shapes of lakes. In addition, modern seismic data show that it is a relatively active tectonic zone.

(3) Supilin Fold-Fault Tectonic Zone

This zone is composed of the N-W-W-trending mountains and some depression troughs, such as the Kyzinengyiche Mountain and Pilin Mountain, between the Karamilan depression and the Kumukuli basin.

Carboniferous clastic rocks and carbonate rocks are dominant, and there are Devonian clastic rocks in the north and Permian carbonate rocks in the south. Mesozoic and Cenozoic terrestrial clastic rocks are confined to the depression troughs. Their tectonic traces are generally distributed in a gentle reverse S-shape, showing a large compound syncline tectonic. Due to the oblique connection of the center of the Supilin fold fault tectonic zone and the Kumukuli fold zone in the Kunlun compound tectonic zone, it is clearly divided into a northern and a southern section. This division not only has a profound impact on the tectonic morphology of the southern and northern sections but also has a certain impact on the strong activities of the tectonic zone during the generation and development processes. The compound syncline tectonic in the zone is composed of many synclines and anticlines, and their axial distribution is deflected in harmony with the overall distribution based on their different tectonic positions. However, there are large differences in fold tectonics at different generation stages and in different locations. The northern section of the tectonic zone is dominated by the middle–upper Carboniferous system. The two wings are compound synclines composed of Devonian and lower Carboniferous systems. The two wings are generally steep, most of their secondary folds are linear, and their axial directions are consistent with the overall trend, but some secondary folds have reversed axial surfaces. In addition, a few folds also have short axes. Therefore, the overall shape is quite complex. In the southern region, with the Triassic system as the core and the Carboniferous system as the western wing, it is generally a simple large syncline tectonic. The folds are composed of Mesozoic short-axis, asymmetric, gently sloping strata, so they are rarely damaged by later fault activities, and the tectonic shape is intact.

6.1.5 *Southwest Tarim–Kumukuli Depression Zone*

(1) Southwest Tarim Depression Zone

This depression zone is located southwest of the Tarim Basin and is a N-W-trending depression zone. It formed in the middle and Late Hercynian and subsided significantly in the Mesozoic and Cenozoic, forming a deep depression zone that is deep in the northwest and shallow in the southeast. The Paleozoic, Mesozoic and Cenozoic systems are well developed and thick and constitute the main oil and gas resource region in the Tarim Basin.

(2) Kumukuli Basin Zone

This basin zone starts from south of Tokuzidaban Mountain in the west and extends to both sides of Qimantag in the east, with a length of approximately 500 km from east to west and a maximum width of nearly 100 km from north to south. Now, it appears as an intermountain basin. Due to the connection and compound of the Supilin fault tectonic zone, it is divided into two sections: the western section is the Shukule Basin (Kalamiran Depression) and the eastern section is the Kumukuri Basin. Both are quite different in scale and triangular, with their shape controlled by the surrounding tectonic system. Except for a small amount of Carboniferous, Permian and Late Hercynian intrusive rocks in the basin exposed in an isolated island-like remnant mountain form, other vast regions are covered by Mesozoic and Cenozoic deposits.

The basin has a narrow deposit range, and only a small amount of Cretaceous and Paleogene–Neogene systems are exposed on the margin. Based on the features of the nearly E-W-trending Carboniferous folds in the central part of the basin and taking into account the influence of the surrounding tectonic systems, it is speculated that the main body should belong to the Kunlun compound tectonic zone and that there are tectonic traces on the margin of the basin that are consistent with the Altyn and N-W-trending systems. The Miocene and Pliocene clastic rock formations in the Kumukuri Basin are widely distributed. Under the influence of the Himalayan movements, a series of nearly E-W-trending anticlines and syncline tectonics formed, accompanied by a small number of same-trending faults on both sides. The fold tectonics generally have lengths of 10–15 km, and a few larger ones have lengths of tens of kilometers. The former extends in the same direction, while the latter is bent and changed in the axial direction. The rock formations on the two wings of the anticline are steep in the north and gentle in the south.[16]

The Miocene–Pliocene deposits in the basin have a thickness of over 5,000 m, which is larger than that of the Shukule Basin in the west. The modern lakes are located on the northern and southern sides, reflecting the movement tendency of the deposit center to both sides (Fig. 6.2).

6.2 N-W-Trending Tectonic System in Central Asia

This tectonic system is located southeast of central Asia and northwest of the Junggar Basin. Three N-W-trending fault zones and matching Mesozoic and Cenozoic basins (namely, the Turgai Basin, Chu-Ayasu Basin and Kyzyl-Kum Basin) are developed.

6.3 N-W-Trending Tectonic System in the North Caucasus

This tectonic system is located in the North Caucasus in southwestern Russia, where the N-W-trending Karbinsk uplift and the North Caucasus basin are developed, and the Greater Caucasus fold zone is located to the south. These uplifts, basins and fold zones formed in the Late Generation and developed into Mesozoic and Cenozoic basins and fault systems.

6.4 N-W-Trending Zagros Tectonic System in the Middle East

This tectonic system spreads north of the Persian Gulf in the middle East, with the Zagros orogenic zone as its main body. It was generated mainly in the late Paleozoic and Mesozoic and is composed of the Zagros fold zone and depression zone, the Arabian slope zone, the Persian Gulf and Persian Gulf basins, the Arabian Basin and the Red Sea. It is generally N-W-trending and is a large N-W-trending tectonic system.

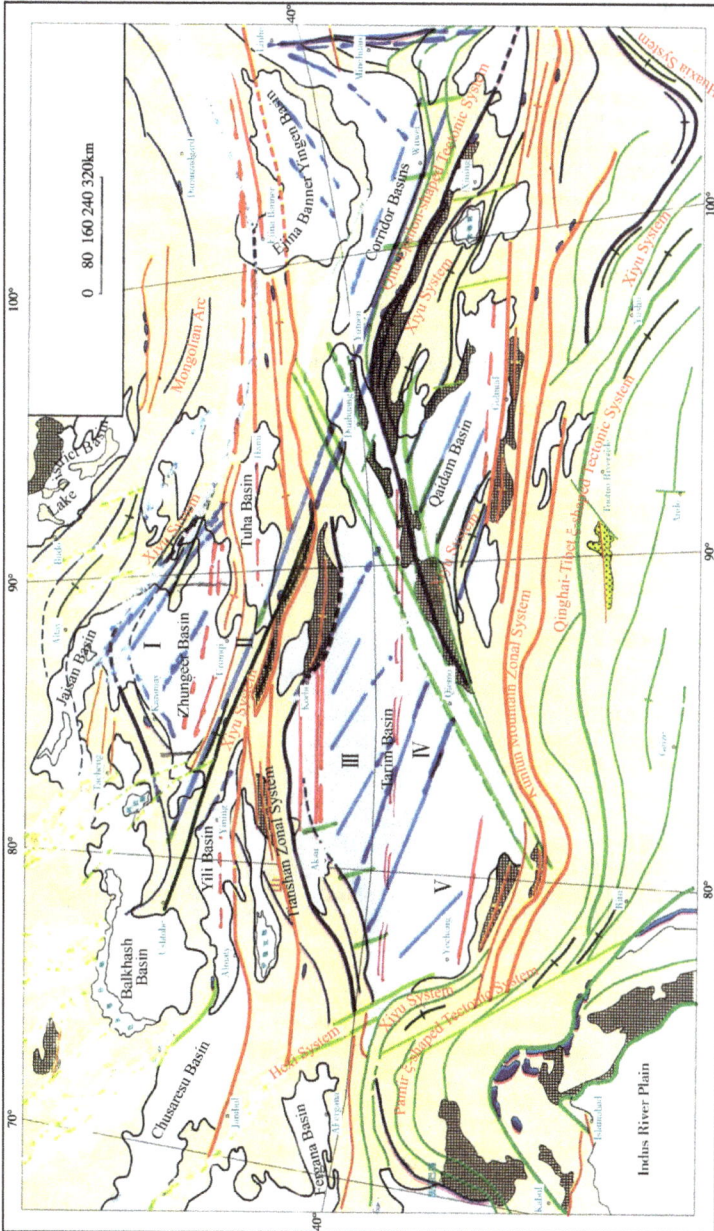

Fig. 6.2.   Schematic diagram of the N-W-trending tectonic systems in northwest China.

Notes: (I) East Junggar N-W-trending tectonic system; (II) Boroholo N-W-trending tectonic system; (III) Awati-Mangar–Qaidam subsidence zone; (IV) Bachu–Qimantage–Suzhilin N-W-trending tectonic system; (V) Southwest Tarim–Kumukuli subsidence zone.

6.5 N-W-Trending Tectonic System on the Western Coast of North America

The tectonic system is composed of a N-W-trending fold zone and a number of Mesozoic and Cenozoic basins from north to south, including Alaska, Alexandria Islands, West Washington, Sacratoro Symphyllum, California, Northeast Basin, Tampico and Velakups Pacific Coast, among others, and is accompanied by multiple compression and compression-torsion fault zones and magmatic rocks, forming a very obvious N-W-trending tectonic system (Fig. 6.3).[17]

6.6 N-W-Trending Tectonic System in the Suez Gulf

The tectonic system is located in the middle of the Suez Gulf, between the Red Sea Mountains and the Sinai Mountains on the east, is N-W-trending,

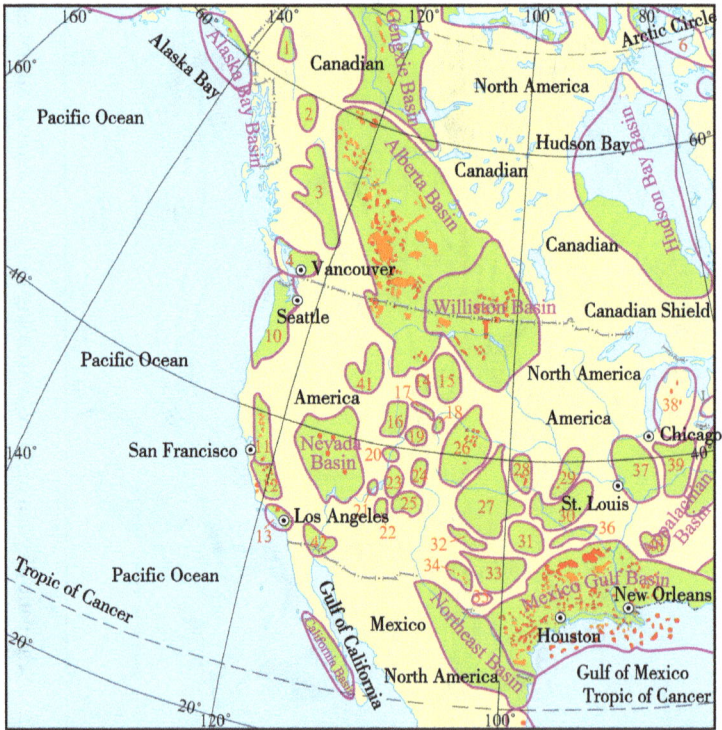

Fig. 6.3. Distribution of N-W-trending tectonic systems and basins on the western coast of North America.

and has a length of 320 km, a width of 50–90 km, and an area of 2.3×10^4 km^2. The deepest sections with a deposition thickness of 7,000 m are located in the sea with a maximum water depth of 80 m. This section also includes a small land section. The Cenozoic deposits are controlled by N-W-trending tectonics.

Similar to the strata in northern Egypt, the strata in this system have an average thickness of 1,000–1,200 m and are composed of Paleozoic–lower Cretaceous sandstones and upper Cretaceous–Eocene carbonate rocks. The rift-stage strata are composed mainly of the Miocene system, with an average thickness of 2,500 m (Fig. 6.4).

6.6.1 Tectonic Features

Due to having different stratigraphic inclinations, the basin can be divided into three sections. The northern strata are inclined to the south-west, the central strata are inclined to the northeast, and the southern strata are inclined to the southwest. The three sections are divided by a 5–7 km transition zone with a relatively gentle inclination. The zone does not completely intersect the entire fault depression, and the sections beyond the transition zone are replaced by complex fault blocks (Fig. 6.5). It was generated in the Late Oligocene and the Early Miocene, and the activities stopped five million years ago (in the Mid-Pliocene). The Miocene–Pleistocene depression was caused by changes in ground temperature.[18]

The Suez Gulf is a nearly N-NW-SSE-trending rift basin, and the southern end of the gulf reaches the Red Sea (Fig. 6.6).

6.6.2 Evolution of the Basin

In the Late Oligocene or Early Miocene, the Suez Gulf Basin first developed as a depression in the Red Sea continental rift system.

The pre-rift deposits began in the Cambrian: the first group of clastic deposits usually refers to the Nubia that may be Paleozoic deposits. The second pre-rift deposits developed mainly in an inherited basin in the Early Cretaceous. It is a set of marine deposits with thicknesses increasing to the north, beginning with deposits of Nubian clastic rocks and carbonate rocks and ending in the Late Eocene.[19]

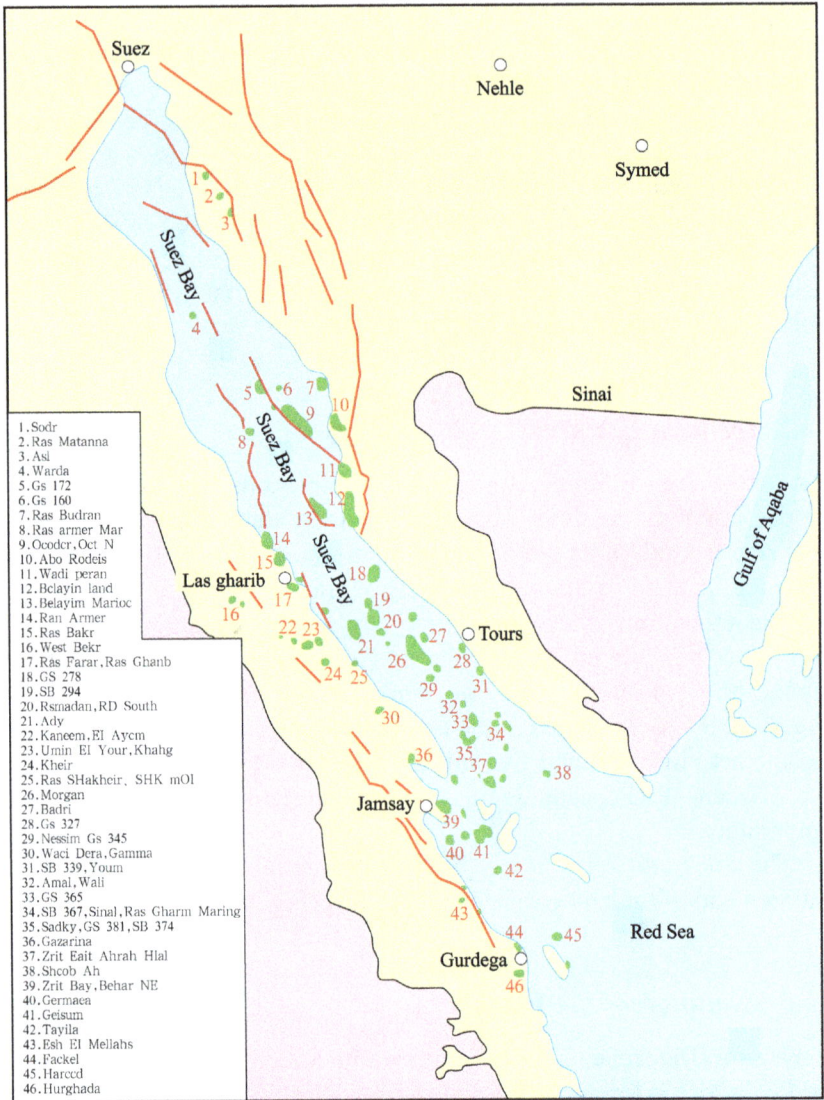

Fig. 6.4.   Distribution of the Neogene Basins in the Suez Gulf.

The Suez Gulf rift began with the opening of the lithosphere in the Late Oligocene and continued to the Miocene. In the initial stage of settlement, there was an uplift on the shoulder of the rift, followed by the entire Red Sea. During the relatively quiet post-rift deposition period of the Suez Bay

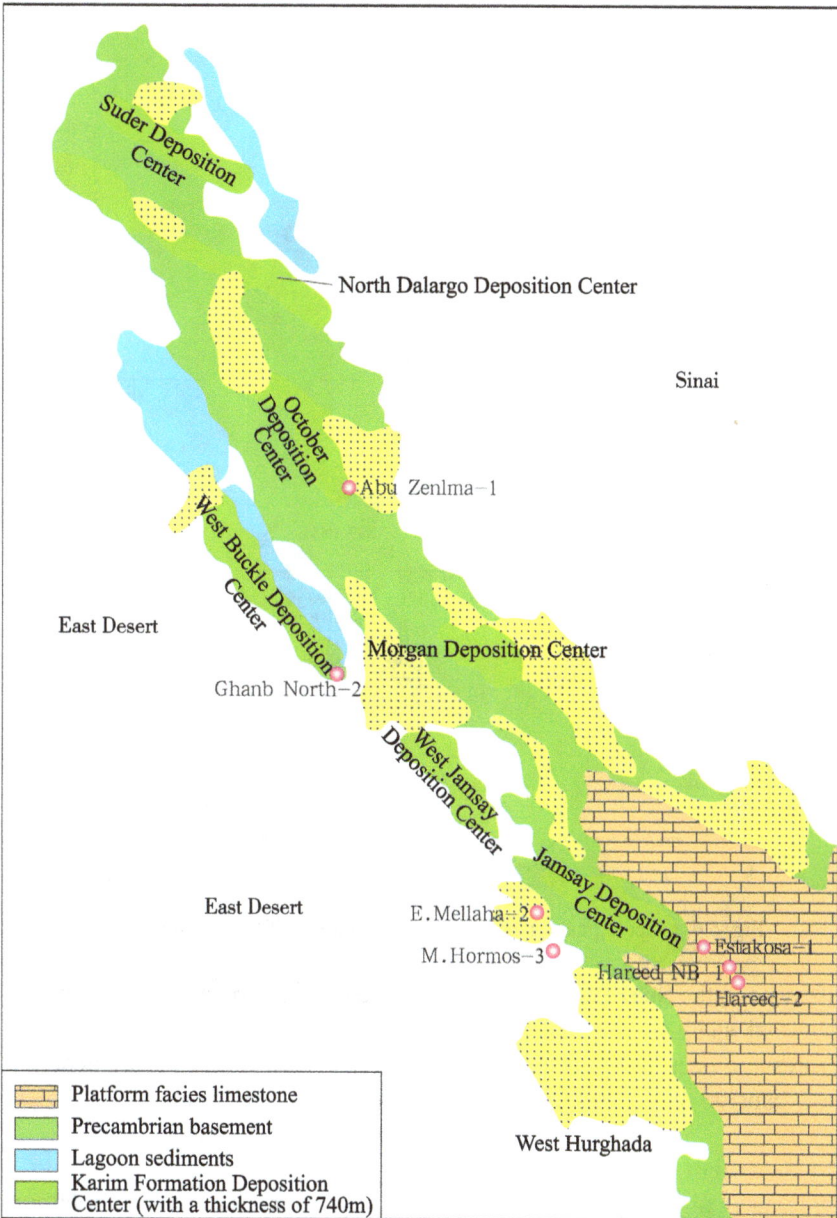

Fig. 6.5.   Tectonic outline of the Suez Gulf basin (HIS, 2010).

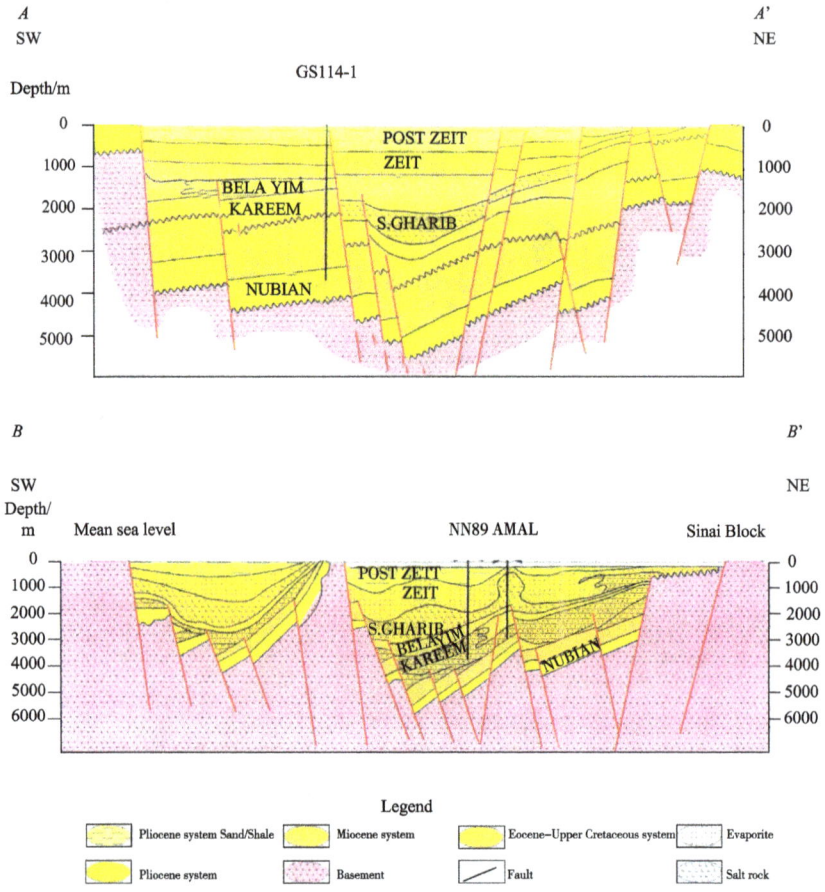

Fig. 6.6.   Geological cross-section of the Suez Gulf Basin (HIS, 2010).

rift system, the Suez Bay basin was deposited mainly by evaporite rocks. In the final stage of the basin's evolution, a continuous geothermal-driven depression occurred in most of the basin, and there was clear evidence of re-expansion on the southern and northern ends of the basin.

(1) Basement
The basement compound includes a variety of crystalline and metamorphic rock types, including schist, gneiss and porphyry with granite properties. In addition, there are a large number of dikes and bedrocks formed by basement intrusive rocks.

(2) Early Post-Rift Depression Unit

This stage featured post-rift thermal depression during the relatively static period of the entire Red Sea rift system. The main tectonics are the reversal and adjustment types and the compaction/deposit-driven types on the resurrection fault. After the Kareem was deposited, local uniformity may have occurred on top of the Kareem Formation in another relatively uplifting period. Afterward, the main controlling role was caused by the post-rift thermal depression, which occurred in the relatively static period of the entire Red Sea rift system, and the Red Sea rift system was active again at the end of the Miocene. This relative uplift formed a barrier that separated the Suez Rift from the previous opening to the Mediterranean Sea. This barrier marked the end of normal marine conditions and the beginning of the first stage of large-scale evaporite deposition in the gulf. However, in the first phase of the evaporite unit, there were two phases of normal marine deposits in the Belayim Formation. The latter generated an important carbonate reservoir and formed an algae formation on the inclined pre-Miocene fault block. The post-Belayim deposits are composed of two larger-scale evaporite rock units, namely, the south Gharib Formation (rock salt) and the Zeit Formation (anhydrite and clastic rocks interbedded with a small amount of rock salt).[20] These evaporative rocks are important caprocks in the basin.

(3) Late Post-Rift Depression Unit

The feature in this stage is the depression caused by the temperature decrease in the formation after active faulting effects and the extensional faulting effects resurrected south of the basin. The main tectonics include half graben, drape tectonic, boundary fault reactivation and reformation of existing fault block traps.

Before the new settlement and accumulation of Pliocene unified Pleistocene clastic rocks started, there was an obvious unconformity on the top of the Zeit Formation. The new deposits appear to reflect the restoration of the extensions south of the Red Sea rift system and the seafloor. Although the Red Sea expanded in line with the Dead Sea–Aqaba transition movements, there is still some evidence that new extensions have occurred in the northern and southern areas of the Suez Gulf since approximately 5 Ma. In the south, a large number of faults intersect the seafloor, and seismic data show that there are still active fault movements. In the north, after the microdeposition of the South Gharib Formation, the thick Zeit Formation and the Post Zeit Formation (EL Tor Group) became adjacent to the Darag Fault, reflecting the effects of the new extension.

References

1. Yuzhu Kang, Cai Xiyuan *et al. Formation Conditions and Distribution of Paleozoic Marine Oil and Gas Fields in China.* Urumchi: Xinjiang Science and Technology Health Publishing House, 2002.
2. Yuzhu Kang, Zongxiu Wang, Zhihong Kang *et al. Research on the Oil Control Effects of the Structural System in the Qaidam Basin.* Beijing: Geological Publishing House, 2010.
3. Yuzhu Kang, Zongxiu Wang, Zhihong Kang *et al. Research on the Oil Control Effects of the Structural System of the Junggar–Tuha Basin.* Beijing: Geological Publishing House, 2011.
4. Liu Jichen. The Continental Block Structure of the North Qilian Orogenic Belt. *Earth Science,* 1991, (6): 635–642.
5. Liu Zerong, Wang Xiaoling. Re-discussions on Hebei–Shandong Brush Structural System. *Journal of the East China Petroleum Institute,* 1981, (2): 1–14.
6. Liu Zenggan *et al. Tectonics and Evolution of the Qinghai–Tibet Plateau.* Beijing: Geological Publishing House, 1990.
7. Anhui Bureau of Geology and Mineral Exploration. *Regional Geological Records of Anhui Province.* Beijing: Geological Publishing House, 1987.
8. Chen Yusui. *Summary of Structural Characteristics of Neocathaysian System in Guizhou Province.* Institute of Geomechanics, Ministry of Geology and Mineral Resources. Collection of Studies on Provincial Tectonic Systems in China (Volume 2). Beijing: Geological Publishing House, 1985.
9. Chen Qingxuan. *Some Noticeable Problems in the Analysis of Rock Deformation and Tectonic Stress Field.* Editorial Board of Journal of Institute of Geology. Journal of Institute of Geology (#8). Beijing: Geological Publishing House, 1986.
10. Cui Shenqing *et al.* Study on Paleotectonic System of Yanliao and Adjacent Areas. *Acta Geologica Sinica,* 1977, (2): 149–162.
11. Mapping Group of Institute of Geology. *Main Tectonic Systems in China.* Beijing: Geological Publishing House, 1978.
12. Institute of Geomechanics. *1:2,500,000 Brief Description of Structural System Diagram of the People's Republic of China and Adjacent Sea Areas.* Beijing: China Cartographic Publishing House, 1984.
13. Dong Shenbao *et al. Metamorphism in China and its Relationship with Crustal Evolution.* Beijing: Geological Publishing House, 1986.

14. Dong Shuwen *et al. Preliminary Study on the Movement Patterns of Dabie Mountain Block.* Editorial Board of Journal of Institute of Geology. Journal of Institute of Geology (#12). Beijing: Geological Publishing House, 1989.

15. Brognon, G., Masson, P. Salt tectonics of the Cuanza Basin, Angola, Portuguese West Africa. *AAPG Bulletin,* 1965, 49(3): 335–336.

16. Bruce, J. *et al.* Increasing River Discharge to the Arctic Ocean. *Science,* 2002, 298: 2171–2173.

17. Camiille, P, Gary, Y. A Globally Coherent Fingerprint of Climatic Change Impacts Across Natural Systems. *Nature,* 2003, 421: 37–42.

18. Biswas, S. K., Deshpande, S. V. Geology and Hydrocarbon Prospects of Kutch, Saurashtra and Narmada Basins. *Petroleum Asia Journal,* 1983, 11: 111–126.

19. Bodard, J. M., Wall, V. J. *Sandstone Porosity Patterns in the Latrobe Group, Offshore Gippsland Basin.* 2nd South-Eastern Australia Oil Exploration Symposium, Technical papers presented at Petroleum Exploration Society of Australia, 1986: 137–154.

20. Fujian Provincial Bureau of Geology and Mineral Resources. *Regional Geological Records of Fujian Province.* Beijing: Geological Publishing House, 1985.

21. Xia Linqi *et al.* A Study of Volcanic Rocks in Orogenic Belts. *Acta Petrologica Et Mineralogica,* 2001, 20(3): 225–231.

22. Song Zhongbao *et al.* Isotopic Age of Shijuli Gabbro in North Mountain and Its Geological Significance. *Acta Geoscientica Sinica,* 2007, 28(1): 7–10.

23. Li Xiangmin *et al.* A LA-ICP-MS Chronological Study of Basic Volcanica in Baiyin Orefield. *Geological Bulletin of China,* 2009, 28(7): 901–906.

Chapter 7

Epsilon-Shaped Tectonic Systems

Yuzhu Kang, Shuwen Xing, Zhihong Kang, Yinsheng Ma, Dewu Qiao, Zongxiu Wang, and Zhihu Ling

Abstract

In this chapter, epsilon-shaped tectonic systems are introduced, including those in China, Eurasia, Irkutsk in southern Siberia of Russia, Teli in Turkey, Gadomein France, England, North America, Cincinnati in southern North America and Brazil in South America.

Keywords: Tectonic system; Epsilon-shaped; Type

The epsilon-shaped tectonic system, proposed by Li Siguang in 1929, is composed of the following parts.

The front arc, or frontal arc, is often an arc-shaped tectonic formed by a number of mutually parallel compression zones, high-angle thrust zones, etc. as the backbone. Within the Northern Hemisphere, this arc generally protrudes to the south, and it protrudes to the west in only a few cases. For the convenience of description, it can be divided into several parts. The middle or front of the arc is called the arc top, and the parts at both ends of the front arc that continue to extend backward are called the two wings.[1] In most cases, the arc top shows the largest curvature, while the two wings show the smallest curvatures. However, there are also some front arcs; the curvature of their top and two wings are not very different, and they combine to form a crescent shape (Fig. 7.1).

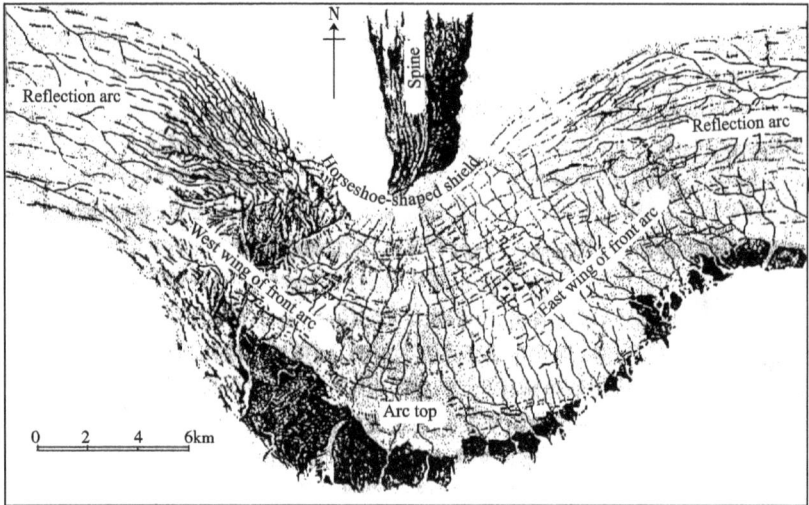

Fig. 7.1.   A small-scale epsilon-shaped tectonic model (based on the surface image from aerial photography of the relevant region).

On the top of the arc-shaped compression zone, the faults are some-times larger in scale, and the strata affected by them may be deeper, result-ing in the arc top sinking into a graben and being covered by new deposits. The folds and thrust faults at the arc top sometimes show violent horizontal compression, forming many similar arc-shaped tectonics. Sometimes the arc-shaped folds formed are not very wide, and the quantity is small. The compression zones on the wings, including the uplift zones of the ancient bedrocks and long basins filled with newer deposits, are approximately parallel to each other or are sometimes arranged in a geese shape. The entire compression zones forming the two wings, such as folds, thrust faults, long basins, etc., increase in number as they extend backward, and the shape expands and spreads out. The reflection arc is located somewhere between the two wings of the front arc, and the arc begins to show a tendency to reverse its bending direction; that is, the two wings tend to spread outwards and bend gradually in the opposite direction to the front part of the front arc and are continuously spreading to the ending sections of the arc, forming two reflection arcs. When the front arc protrudes to the south, the reflection arc protrudes to the north. When the front arc protrudes to the west, the reflection arc protrudes to the east. Sometimes, the reflection arc is no less than the front arc in scale, while sometimes, the front arc is smaller in scale and curvature, and other times, it only slightly bends outward. The arcs are not crowded in several narrow regions, such as the main part of the front arc, but are scattered over a relatively wide region.

Notably, there is no boundary between the arc top, the two wings and the two reflection arcs of an epsilon-shaped tectonic. Together, they are slightly sinusoidal, that is, one side is S-shaped, and the other side is reverse S-shaped. They are united at the top of the front arc to form a continuous, repeatedly curved compound tectonic zone.[2] This joining, however, does not mean that the zone's various tectonic zones are completely continuous from one end to the other end of the reflection arc.

The spine is located in the depression region of the front arc, which means that the middle region is half-surrounded by the front arc, and there is often a strong linear uplift compression zone. In a few cases, this uplift compression zone may be subject to a settlement (or is *quasi*-trough-like) before its uplift. In some special cases, after the uplift, whether it is possible to become a trough zone in a depression is still an open question. The position of this uplift compression zone is approximately the same as the bilateral symmetry axis of the front arc, which is the spine of the epsilon-shaped tectonics. These complex compressional tectonic zones formed by several compressional zones are generally limited to a certain range, but sometimes, they are scattered. Among them, the region with the most severe compression mostly faces the top of the front arc, and it is approximately at right angles to the top of the front arc. On both sides of this strong compression zone, there are often weaker compression zones. The farther these compression zones are from the central strong compression zone, the weaker they are, even disappearing. The compression zone that constitutes the entire spine becomes weaker as it approaches the arc top and finally disappears completely at a certain distance from the arc top. The abovementioned compression zone is composed of folds, thrust surfaces, compression crushing zones, splitting surfaces, fragmentation and foliation. In the direction at a right angle to the compression zone, there are often tensional faults or normal faults.

What is described above is the normal form of the spine. Whether the spine can also appear as a wide uplift, that is, a small range of wide folds, depressions or compound depression zones (*quasi*-geochannels) is still an open question. Because there are often ancient rocks exposed within the range around the spine, the compression zone that forms the spine is often compounded on the older compression zone. Of course, the compression directions of these older tectonics are not necessarily the same as those of the spines of the epsilon-shaped tectonics. When the spine of an epsilon-shaped tectonic is formed, it must be an uplift zone due to compression. However, if this uplift zone is later subjected to tension in the direction at a right angle to its axis, some large faults may occur near or on both sides of it, forming a graben, as mentioned above.

The horseshoe-shaped shield is located between the spine and the top and the two wings of the front arc, and there is often a horseshoe-shaped flat region or a region with extremely weak folds. When the front curvature of the epsilon-shaped tectonics is not large, a vast and flat shield often forms. This shield, as a component of the epsilon-shaped tectonic, may be composed of ancient folds, faults, or other rigid tectonic features, and there may also be new folds, faults, or other tectonic features through the shield. Of course, all these old and new tectonic traces do not belong to the epsilon-shaped tectonic system. Therefore, their existence does not affect the formation stability of the horseshoe-shaped shield.[3] However, when the front curvature is very large, this horseshoe-shaped region is still unavoidably affected by some relatively weak and minor-axis folds. All or part of some horseshoe-shaped shields, as well as their ground, are composed of ancient blocks with folds or faults, and all or part of some other horseshoe-shaped shields are covered with a certain thickness of flat rock formation on the basement of the ancient folds and faults. The former is sometimes called a platform, and the latter is sometimes called a basin. However, in terms of the overall epsilon-shaped tectonics, this kind of basin on one side of the spine and the land blocks with ancient faults and folds on the other side have equal importance.

In addition to the main components of the abovementioned groups of epsilon-shaped tectonics, there are sometimes some secondary tectonic traces in the depression region of the reflection arc, but in terms of geomechanics, they are not necessarily secondary. In the recessed region of the reflection arc, there are often horizontal spiral tectonics at varying scales, as well as in the range of the horseshoe shield, especially in the middle of the horseshoe shield and the section near the front arc. The reflection arc depression region is relatively stable and may form a basin or platform, but sometimes, there are quite severe folds in its middle that form the spine of the reflection arc. In this case, the relatively stable regions on both sides also become small horseshoe-shaped shields.

In front of the front arc top, due to strong tension and cracking effects, sometimes the granite masses are exposed or buried in shallow underground positions as well as on the top of the reflection arc.[4]

The N-S-trending epsilon-shaped tectonic spine is sometimes generated in the N-S-trending compressional region, including N-S-trending synclines and anticlines. In addition, because the epsilon-shaped tectonic spine has already been generated, the E-W-trending compression can be triggered in the following tectonic movements, forming a composite phenomenon of the epsilon-shaped tectonic spine and a N-S-trending tectonic zone that does not belong to the epsilon-shaped tectonic system. The latter are found more frequently in China. If they are generated behind the front arc of the epsilon-shaped tectonic, especially in the middle behind the front arc, they will inevitably be confused with the components that make up the epsilon-shaped tectonic spine, but they are not indistinguishable. As mentioned, the main difference between them concerns their spread or distribution. The pure N-S-trending tectonic zone often passes through the front arc of the epsilon-shaped tectonic system, while the N-S-trending tectonic zone belonging to the spine can never pass through the front arc. The pure N-S-trending tectonic zones are often strictly parallel to each other and spread over a wide range, while the various tectonic zones that make up the epsilon-shaped spine are densely packed in the middle region behind the front arc and are often narrowed frontward (toward the arc top) and widened backward (away from the arc top).

In general, except for some abnormalities in a few cases due to distortion or damage, the main components of all epsilon-shaped tectonics are arranged symmetrically on both sides with the spine as the axis and with the two wings as the horns to each other, forming a whole with the above-mentioned morphological features. The tectonic elements that make up its various components, such as folds and faults, are also arranged in certain rules or intersect with each other. These arrangement rules have a certain control effect on the distribution of minerals, especially the enrichment zones of mineral deposits in the vicinity of the front arc and the reflection arc with the maximum curvature. In regions with epsilon-shaped tectonics, the above rules can be used as guidance for exploration plans and construction designs, which is very important.

From the rules of these tectonic morphologies, some more important facts have been discovered. In China, the front arc of the epsilon-shaped tectonic system generally protrudes to the south, while the front arc of only a few tectonic systems protrudes to the west. Some epsilon-shaped tectonics determined in other regions of the Northern Hemisphere are also arranged according to the same rules. The orientation of the epsilon-shaped tectonics is amazing in geological tectonics. It clearly indicates that the origin of these tectonic systems is closely related to the orientation of the Earth's axis of rotation, such as the E-W-trending compound tectonic zones and the N-S-trending tectonic zones.

In a deformation image of epsilon-shaped tectonics based on theoretical analysis and simulation experiments, the regions involved in these tectonics can be treated as a flat slab.[5] The load applied on this flat slab is uniform and horizontal. The application direction of the load is generally from high to low latitude or in some cases, from east to west. When this flat slab and the underlying rock masses are tightly fixed only not far from the two ends of the flat slab, and when other sections are easy to slide or twist, they will bend slightly in the direction of the applied load. The middle of the beam, namely, the position equivalent to the top of the front arc, is obviously bent, and the spread, arrangement and inter-intersection of the epsilon-shaped tectonic lines, especially the compression and tensional tectonic lines, also reflect the shape of the principal stress traces occurring in the flat slab. In the position in which the flat slab and its underlying rock masses (may be the so-called basement or the deeper layers) are fixed tightly, the basement is generally consistent with that in the region with a stable reflection arc concave surface.

The depth of the epsilon-shaped tectonics is still uncertain. However, generally speaking, the smaller the thickness of the rock strata affected by the smaller epsilon-shaped tectonics is, the larger the scale is, and the greater the thickness of the affected rock strata is. Small and medium-sized tectonic systems have not been found to belong to this category. Among the epsilon-shaped tectonics discovered, the smallest one measures over 30 km long from the end of one reflection arc to the end of the other reflection arc and over 20 km from the top of the outermost front arc to the furthest spine point away from the front arc. The scale of this type of tectonics is still uncertain.

7.1 Epsilon-Shaped Tectonic System in China

Most epsilon-shaped tectonics discovered in China developed after the Triassic, and some of them were active in the Tertiar. Since they have been developed in this way since the Mesozoic, is it possible that there are no epsilon-shaped tectonics in the older tectonic layers? Obviously, older tectonic systems are more difficult to identify because most of them have been destroyed or covered, and the ancient epsilon-shaped tectonics that may exist are not necessarily consistent with the epsilon-shaped tectonics that can be clearly identified on the Earth's surface today. In other words, there is still no basis for speculating the possible forms and spread ranges of ancient epsilon-shaped tectonics with the spread of epsilon-shaped tectonics generated since the Mesozoic.[6]

In the past few years, a number of epsilon-shaped tectonics have been identified in China. The respective features of these epsilon-shaped tectonics and their respective interferences or their compound relationships with other tectonic systems will be discussed later.

Each of the identified epsilon-shaped tectonics has been understood through field survey works and practices in a certain period.

In addition, some sections of a number of tectonic systems have the same features of epsilon-shaped tectonics, but further investigation is needed to identify the tectonic types of the remaining sections.

Since so many epsilon-shaped tectonics have been discovered in China, they must exist in some regions of the Northern Hemisphere outside of China.

The generation process of the epsilon-shaped tectonic system can be approximately divided into three stages, namely, the embryonic stage (or infant stage), the forming stage (or mature stage) and the deformation stage (or old age), which approximately correspond to the budding stage, the forming stage and the complication stage, respectively, as divided by Zhang Guoduo.

(1) *Embryonic stage (or infant stage)*: This stage is the starting stage of the epsilon-shaped tectonic system. In this stage, the front arc begins to appear and manifests as a front arc depression zone or an early deformation zone. Some of the arc-shaped tectonic zones determined thus far might be in the embryonic stage of the epsilon-shaped tectonic system.

(2) *Forming stage (or mature stage)*: This stage is the violently active stage of the epsilon-shaped tectonic system. In this stage, activities are violent, forming the basic outline of the epsilon-shaped tectonic system and obvious deformation zones. The front arc is relatively open, and the arc top is round and smooth, composed mostly of folds (normal folds) in the early period. Thrust faults gradually appear, and the spine is generated in the later period. The early period is very short and far from the front arc, which is often composed of normal folds, and gradually expands in the later period, gradually forming the front arc and thrust faults.

The shield is quite extended in the early period, with nearly no tectonic traces, and gradually shrinks in the later stage, with slight folds. Most epsilon-shaped tectonic systems determined thus far have undergone the forming stage.

(3) *Deformation period (or old age)*: This period is the remaining activity period of the epsilon-shaped tectonic system. Activities gradually reduce, the epsilon-shaped tectonic system deforms, and gradually deviates from the normal form: the front arc shows a larger curvature, the wing angle decreases, the arc top sharpens and even becomes U-shaped or V-shaped, the spine expands further forward and is closer to the front arc, its front top might be divided into several short ridges, its ends might be bent or branched, and the shield might become narrower or disappear. In some epsilon-shaped tectonics, not only did the late front arc regenerate within the spine, but the late spine also regenerated inside of the late front arc. Both are compounded within the system to form an epsilon-shaped tectonic structure. Only a few of the epsilon-shaped shapes determined thus far are in the deformation stage.

However, some epsilon-shaped tectonics may suffer interference from other tectonic systems during the generation process, resulting in congenital deficiencies. After some epsilon-shaped tectonics are generated, they may be transformed by other tectonic systems, resulting in late destruction. Therefore, it is not required that all the components of the epsilon-shaped tectonics are complete. In practice, as long as the front arc and the spine are well developed, an epsilon-shaped tectonic system can be established. There are two major difficulties or problems in the determination of the epsilon-shaped tectonics. (a) The first concerns whether the front arc exists and whether it is a complete arc-shaped tectonic zone or two cross folds or faults. (b) The second concerns whether the spine exists and whether it and the front arc are a unified whole that is generated under the same stress field and have a genetic connection or they are two parts that are interfered and compounded with each other under two different stress fields. If the front arc is well developed, with a strong continuity,[7] it should be not difficult to identify. If the front arc top is broken and covered by newer deposits, emplaced by later magmatic rocks, or interfered with by other tectonic systems, it is likely to be more difficult to identify. Therefore, in the identification, more attention should be given to whether the formation and deformation features of the two wings of the front arc have certain commonalities and continuity to determine whether they have arc features. There are more problems in the identification of spines. Many epsilon-shaped tectonic systems are determined by the front arc instead of the spine. For example, when the Huaiyang arc was first identified as an epsilon-shaped tectonic front arc in 1933, the Huaiyang Mountain from Lie Mountain north of Xiao County was regarded as the proposed epsilon-shaped tectonic spine, which was very reluctant. In 1946, Sun Dianqing and Xu Yujian discovered that the 40-mile Chang Mountain N-S-trending folds and thrust faults between the Gugu in Henan and the Huoqiu in Anhui extending are truly a Huaiyang epsilon-shaped tectonic spine, and are also considered as not commensurate with the powerful Huaiyang arc. Thus, it is hypothesized that it may have been caused by the depression of the region. In 1954, Pro. Li Siguang speculated that the Huaiyang epsilon-shaped tectonic spine might extend southward continuously along the 40-mile Chang Mountain zone and was compounded with other tectonic systems in the Proterozoic–Archaean metamorphic rock system in the Dabie Mountains. Wu Leibo, Ning chongzhi *et al.* carried out a systematic survey of the Dabie Mountain region in 1955 and 1958 and finally determined that the Huaiyang epsilon-shaped spine was located in Gugu, Huoqiu, Jinzhai, Huoshan, Luotian, Yingshan and other places on the border regions in Henan, Anhui, and Hubei. In other words, 25 years passed from determining the front arc to determining the spine. For Irkutsk epsilon-shaped tectonics, in 1929, Pro. Li Siguang pointed out that there might be a large epsilon-shaped tectonic in the Irkutsk region based on the existence of the Sayan–Baikal arc, speculated that its spine should be located east of the Angara River, and prepared a drawing in the form of a dashed line in the *Geology of China* (published in 1939). The correctness of this inference was unintentionally verified by the Soviet geologist Mordowski in 1952 when he discovered a N-S-trending fold fault zone east of the Angara River.

In Mainland China, more than 60 epsilon-shaped tectonic systems have been determined or proposed and are described as follows.

7.1.1 *Qilu–Helan Epsilon-Shaped Tectonic System*

The Qilu–Helan epsilon-shaped tectonic system is located in the middle and north of China, passing through Xinjiang, Qinghai, Gansu, Ningxia, Shaanxi, Shanxi, Hebei and Beijing and other provinces, municipalities and autonomous regions. It is sandwiched between the Tianshan–Yinshan E-W-trending tectonic system and the Kunlun–Qinling Mountains

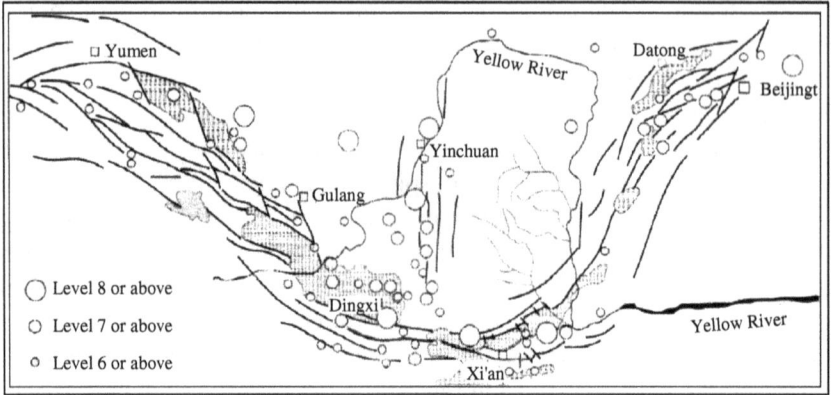

Fig. 7.2.   Relationship between the Qilu–Helan epsilon-shaped tectonics and the earthquake epicenter distribution.

E-W-trending tectonic system. Its geographical location is at 92°00'120°00'E and 34°00'–41°00N, with a length of 2,000 km from east to west and a width of 900 km from north to south. It is a large epsilon-shaped tectonic system (Fig. 7.2).

(1) Scope and Composition of the Qilu–Helan Epsilon-Shaped Tectonic System

This epsilon-shaped front arc spreads in the Qilian Mountains, Longshou Mountains, Shulenan Mountains, Laji Mountains, Luliang Mountains, Wutai Mountains and Heng Mountains. It is an E-W-trending arc-shaped tectonic zone protruding to the south, referred to as the Qilu Arc. The top of the front arc is located near Baoji. West of Baoji, the tectonic line gradually turns from E-W-trending to N-W-trending. The most notable tectonic traces include the Huajialing–Baoji anticline, the Tianshui–Wushan fault zone and the Lixian–Tongren fault zone.[8] East of Baoji, the tectonic line gradually turns from E-W-trending to N-E-trending. The most notable tectonic activity is the Fenwei graben, followed by the Tongchuan compound anticline and the Zhongtiaoshan compound anti-cline. The arc top is reconnected with the Kunlun–Qinling Mountains E-W-trending tectonic system.

The western wing of the front arc extends in Jiuquan, Minle, Lanzhou and Dingxi, approximately corresponding to Heli Mountain, Longshou Mountains, Maya Snow Mountain, Halagu Mountain, Laji Mountain and

Qilian Mountain. It is presented by the N-W-trending fold zones, fault zones and troughs sandwiched between them in a reverse ξ-shaped arrangement. East of Lanzhou, it is composed mainly of the Huajialing–Baoji Great Anticline, the Tianshui–Wushan fault zone and the western part of the Li County–Tongren fault zone extending from the top of the front arc. Between Lanzhou and Linze, it features anticlines and troughs arranged alternately and parallelly and is accompanied by the same-trending thrust faults, including the Heli Mountain–Longshou Mountain fold zone, Zhangye–Minle trough, Maya Snow Mountain anticline, Menyuan trough, Datong Mountain–Qingshiling anticline, Xining–Minhe trough, Riyue Mountain–Laji Mountain anticline and Xunhua trough from north to south. Between Linze and Jiuquan, it is composed mainly of large anticlines and fault zones reconnected in the older N-W-trending tectonic zone, including the Qilian Mountain Main Peak–Corridor South Mountain compound anticline, upper Heihe trough, Tuolai Ranch trough, Tuolai Mountain compound anticline, upper Datong River trough, upper Shule River trough and Shule South Mountain compound anticline from north to south.

The eastern wing of the front arc extends in Hancheng, Lishi, Ningwu and Datong, approximately corresponding to Luliang Mountain, Wutai Mountain and Heng Mountain. It is presented by the N-E-trending large anticlines (or land ridges) and large synclines (or troughs) in a ξ-shaped arrangement and is accompanied by the same-trending thrust faults, including the Yangyuan anticline, Yangyuan South Mountain fault, Sanggan River trough, Sanggan River South–Nankou anticline, Hunyuan trough, Guangling–Yu County fault, Hengshan anticline, Baihua Mountain syncline, Fanzhi trough, Wutai Mountain–Lvliang Mountain anticline and Taiyuan trough from north to south.

The Qilu–Helan epsilon-shaped tectonic spine spreads in Dengkou, Yinchuan, Zhongning and Pingliang on the northern side of the front arc and is a N-S-trending long tectonic zone protruding to the west in its middle, referred to as the Helan Mountain fold zone. It is reconnected on the Helan Mountain N-S-trending tectonic system and extends to the southern side of the Tianshan–Yinshan E-W-trending tectonic system to the north. The spine is composed of N-S-trending folds and thrust faults, including the Delaiji–Zhongwei fault zone, Helan Mountain anticline, Zhongning–Tongxin anticline, Shizuishan–Yinchuan–Guyuan syncline (or trough), Zhuozishan–Qinglongshan–Pingliang anticline and its eastern fault and Yanchi–Huan County syncline from west to east.

Its shield is distributed on the eastern and western sides of the spine. The Aning (Alxa-Huining) shield is located in the west and was a shield land with a small amount of uplift in the Mesozoic. The Mesozoic system was deposited only in a local region, and it became a relatively large descending region in the Cenozoic. The famous Yishan Shield is located in the east. It was a relatively declining basin over a long period from the Triassic or Jurassic to the Paleogene–Neogene; its inner rock formations are gentle, and only gentle folds or domes appear in the marginal region.[9]

Qilu–Helan Epsilon-shaped tectonic reflection arc: The western wing reflection arc spreads in Jiuquan, Yumen, Sunan, Annanbar and other places. It is an arc-shaped tectonic zone protruding to the north. The arc top is located north of Qiaowan and is composed of NWW-NEE-trending folds and thrust faults, including the Liuyuan fault zone, Yufengshan–Liuyuan–Houhongquan syncline, Qiaowan–Dunhuang fault zone, Anxi–Dunhuang uplift zone, Yumenzhen–Jiayuguan North trough, Subei–Jiayuguan fault zone and Changma–Yumen trough from north to south. The northern section of the reflection arc extends into the southern side of the Tianshan–Yinshan E-W-trending tectonic system, its two wings are connected obliquely with the E-W-trending tectonic system, and the middle section is reconnected with the Tianshan–Beishan E-W-trending tectonic system. The western wing reflecting arc spine spreads in Yanchi Bay, Yueya Lake and its southern region. It is composed mainly of nearly N-S-trending thrust faults and folds and disappears south of the Shule South Mountain fault zone to the north. Its extension to the south is unclear, but there may be some traces north of Ulandaban Mountain. The eastern wing reflection arc spreads in Datong, Xuanhua, Chengde, Qinhuangdao and other places. It is an arc-shaped tectonic zone protruding to the north. The arc top is located near Chengde. It is composed of N-W-W-trending, N-E-E-trending or arc-shaped folds, troughs and thrust faults. The outer arc includes the Zhuolu–Huilai trough, Langshan–Jiuxian fault, Dagucheng–Yongning fault, Yanqing trough, Liping anticline, Wumishan fault zone, Longhua anticline, Chengde syncline, Qinglong–Qinhuangdao anticline and Qian'an–Changli anticline. The inner arc includes the Lixian fold zone, Xiaotangshan–Xiacang fold fault zone and Beijing Xishan fold fault zone. The northern section of the reflection arc extends south of the Tianshan–Yinshan E-W-trending tectonic system, its two wings are connected obliquely with the E-W-trending tectonic system, and the middle section is reconnected with the E-W-trending tectonic system. Geophysical data show that there are some traces of the components of the reflection arc in the Bohai Sea, Lushun and Dalian.

(2) Genetic Age of Qilu–Helan Epsilon-Shaped Tectonic System

Existing data indicate that the Qilu–Helan epsilon-shaped arc tectonic zone was generated in the Carboniferous, and the spine appeared in the Late Triassic. After the Indosinian Movement at the end of the Late Triassic, the basic outline of the epsilon-shaped tectonic zone was roughly formed, the front arc and spine were uplifted continuously, the Yishan shield was depressed continuously, the Aning shield was uplifted slightly, and a series of reverse ξ-shaped troughs appeared on the western wing of the front arc, which provided a place for Jurassic deposits. During the early and middle Yanshan movement from the end of the Late Jurassic to the end of the Early Cretaceous, the arc-shaped tectonic zone and the spine matured successively, and the entire Qilu–Helan epsilon-shaped tectonic system finalized. In the Late Cretaceous, the epsilon-shaped tectonic system was uplifted as a whole. From the Paleogene to the Neogene, the troughs on the western wing of the front arc and the Aning and Yishan shields were deposited again. In the Neoid, the epsilon-shaped tectonic system was still strongly active, and it is one of the most active tectonic systems in central and northern China.

(3) Ore-Controlling and Earthquake-Controlling Effects of the Qilu–Helan Epsilon-Shaped Tectonic System

The ξ-shaped or reverse ξ-shaped troughs in the Qilu–Helan epsilon-shaped arc tectonic zone are generally the regions where mineral deposits are concentrated. For example, the Menyuan trough and Xunhua trough on the western wing of the front arc and the Sanggan River trough, Fanzhi trough and Taiyuan trough on the eastern wing of the front arc are important coal-accumulating basins. The Aning shield and the Yishan shield are also concentrated sites for coal-bearing formations in various periods. In these troughs or shields, the low-order, low-level geese-shaped anticline or torsion tectonic systems formed by the relative torsions are also favorable positions for the accumulation and preservation of oil and natural gas.[10]

The composite sections of the Qilu–Helan epsilon-shaped and other tectonic systems are often enriched by endogenous metal minerals. The Qilu–Helan epsilon-shaped tectonic system is also one of the important seismogenic tectonics in central and northern China, with frequent and strong earthquakes. Many magnitude 7.8 earthquakes are distributed along the front arc and the reflection arc, which is a zone of strong earthquake activity.

7.1.2 *Huaiyang Epsilon-Shaped Tectonic System*

The Huaiyang epsilon-shaped tectonic system is located in the middle and lower reaches of the Yangtze River through Hubei, Henan, Anhui, Jiangsu, Jiangxi and other provinces. It is located between $109°00'E–120°00'E$ and $29°50'N–32°20'N$, with a length of approximately 1,000 km from east to west and a width of 300 km from north to south. It is a large epsilon-shaped tectonic system (Fig. 7.3).

(1) Structure and Main Components of the Huaiyang Epsilon-Shaped Tectonic System

(a) *Front arc*: The front arc spreads along Xiangyang, Yingcheng, Wuhan, Jiujiang, Anqing, Tongling, Nanjing and Zhenjiang and is an E-W-trending arc-shaped tectonic zone protruding to the south, commonly

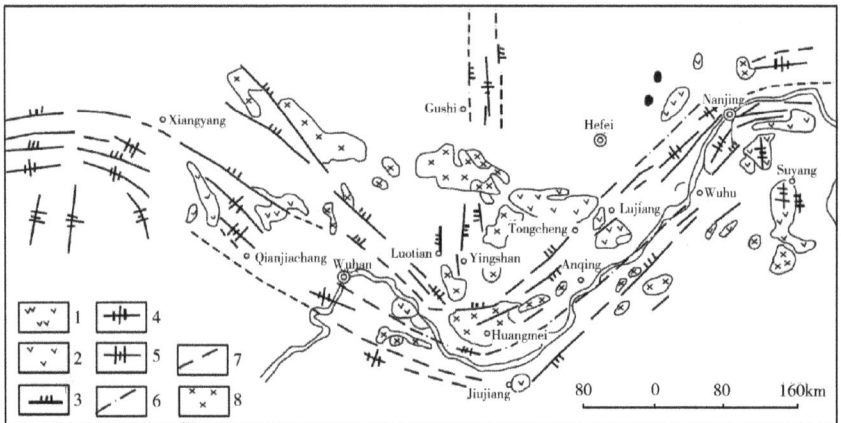

Fig. 7.3. Distribution relationship between Huaiyang epsilon-shaped tectonics and Yanshanian magmatic rocks.

Notes: (1) upper Cretaceous intercalated basalt; (2) upper Jurassic–Lower Cretaceous volcanic rock system; (3) Main fault zone; (4) Anticline axis; (5) Syncline axis; (6) Aeromagnetic presumed fault; (7) Presumed fault; (8) Intermediate-acid intrusive rocks.

known as the Huaiyang arc. It is composed of a series of folds, thrusts and Mesozoic acidic rock zones. It is divided into inner, middle and outer (or northern, middle and southern) arc tectonic zones by four intersecting arc-shaped fault zones.

Inner zone: Its northern boundary is a Chuhe–Tongcheng N-E-trending fault zone on the eastern wing of the fore arc, and it gradually turns N-E-trending and E-W-trending after passing through Taihu Lake. It is connected to the Qichun–Fangfan dynamic metamorphic zone and the fold fault zone on the western wing of the fore arc after passing through the northern section of the Meichuan granite mass and to the Yingshan–Ciyang fault zone on the northwest and then extends to the Nanyang Basin and west of it.[11] There is a N-E-trending dynamic metamorphic zone with a width of more than 7 km in the Chuhe–Tongcheng fault, with trough and intermittent Yanshanian magmatic activity. The Qichun–Fangfan dynamic metamorphic zone has a width from several kilometers to 30 km and a length of 100 km. The mylonitized granite and cataclastic granite masses are formed by the Yanshanian Fangfan and Shuangfeng sharp intrusions. The two dynamic metamorphic zones correspond to the east and west, with the same features and ages. The southern boundary of the inner zone is the Luohe–Wujiang N-E-trending fault zone on the eastern wing of the front arc, and it extends to the eastern wing of the reflection arc on the east, forming the northern boundary of the Ningzhen arc. It extends to Susong and Huangmei to the west, gradually turns nearly E-W-trending, passes through Dajin and Qichun to the west, then gradually turns N-W-trending and connects to the Qizhou–Anlu–Xiangfan fault zone. The two arc-shaped fault zones are composed of several secondary uplift zones, fold fault zones and depression zones. They form a ξ-shaped oblique arrangement on the eastern wing and a reverse ξ-shaped oblique arrangement on the western wing. The Yanshanian intermediate-acidic intrusive rocks and volcanic rocks are distributed intermittently in the zone and are superimposed by the Cretaceous–Paleogene–Neogene basins.

Middle zone: The southern boundary is the N-E-trending Xiaodanyang–Tongling–Wangjiang fault zone on the eastern wing of the front arc. It gradually turns E-W-trending after passing through the Wangjiang River to the west, reaches Daye along the Yangtze River via Wuxue, and is then connected to the Hanyang–Nanzhang tectonic zone, which is also shown on the northeastern margin of the Jianghan Basin. The Upper Paleozoic and Triassic systems are well developed in this zone, with a wide distribution and good preservation. Late Jurassic–Cretaceous intermediate-acid volcanic rocks, subvolcanic rocks and Yanshanian mid-alkaline intrusive rocks are widely developed. The Wuhan–Ningzhen region features the intrusions and eruptions of high-alkali magma. The volcanic rock basins and intrusive rock uplifts are alternately connected to each other, forming a significant magmatic rock zone. West of the Jianghan Basin, this zone features basic volcanic activity. In addition, the arc-shaped aeromagnetic anomaly zone is extremely noticeable along the zone.

Outer zone: The southern boundary is on the eastern wing of the front arc, which is the Nanling–Qingyang–Lushan N-E-trending fault zone, with a large scale and good continuity. It is mostly reconnected or mitered with the N-E-trending tectonic zone and gradually turns nearly E-W-trending in Jiujiang and Yangxin, and the WNW-trending after Sanxi and Shuangxi follows the Dengnan–Qianjiachang fault zone. Both the aeromagnetic anomaly and remote sensing images clearly indicate this zone, and the magmatic rock zone, copper-sulfide zone and physical anomaly zone are also consistent with those in this zone. The Jiujiang–Yangxin rupture deformation is not significant, and it is composed of a series of nearly E-W-trending secondary folds, faults and small thrusts, which may be a ductile compression-torsional tectonic zone.

In addition, the strength and type of folds in the front arc fold fault zone show regular changes with the sequence of tectonic movements. That is, the middle Triassic and previous rock formations are dominated by closed folds, with reverse folds and fan-shaped folds. On the plane, the eastern wing is mostly S-shaped, and the western wing is mostly reverse S-shaped.[12] The Upper Triassic and Unified Middle Jurassic systems are dominated by widespread and short-axis folds and some reverse folds. The upper Jurassic volcanic rocks and clastic rocks are dominated by widespread folds.

The top of the front arc is located east of Wuxue, at approximately 115°40'E. Its northern section is composed mainly of Yanshanian Meichuan granites, showing an arc protruding to the south. The arc turning point is located at Jingzhupu, in which there are a number of E-W-trending Mesoproterozoic Hong'an Group xenoliths with E-W-trending flakes. South of Jingzhupu, a group of nearly N-S-trending quartz porphyry veins are developed, converging to the north and diverging to the south. The E-W-trending Sinian system and the underlying Mesoproterozoic Hong'an Group extend near Wuxue to the south of the arc top. West of the arc top are the N-W-W-trending Tianzhen compound syncline and the thrust faults on the two wings composed of Paleozoic and Triassic systems on the Wuxue–Qinzhou region and the Mesozoic–Cenozoic subside significantly; the area east of the arc top is covered by the Cretaceous–Paleogene–Neogene and Quaternary systems in the Wuxue–Huangmei region. Geophysical prospecting data indicate that there are N-E-E-trending Paleozoic and Triassic folds, which gradually turn E-W-trending to the west. The river network is developed, and the lakes are dotted at the arc top.

(b) *Spine*: The spine spreads in Gushi, Huoqiu, Jinzhai, Huoshan, Macheng, Luotian and Yingshan on the northern side of the front arc and is a N-S-trending tectonic zone, composed mainly of N-S-trending or nearly N-S-trending thrusts and compressional fault zones, as well as the same-trending small folds, schists and vein rock zones. The tectonic traces are relatively scattered but concentrated along certain longitudes. From west to east, they are (a) the Sanlifan zone, with a width of 13–15 km and a length of 75–80 km, dominated by the N-S-trending compressional fault zone and accompanied by thrust and schistosity zones, (b) the Manshui River Zone, with a width of 10–12 km and a length of 60–65 km, dominated by N-S-trending schist zones and thrusts, and the Yingshan zone (the strongest zone), with a width of 15–18 km and a length of approximately 100 km, (c) the N-S-trending compressional fault zone, dominated by dense thrusts, schistosity zones and small folds and accompanied by Yanshanian rock masses and dykes, and (d) the N-E-trending folds on the eastern wing of the arc-shaped tectonic zone and merged by this zone and spread in the N-S-trending direction at 115°40'E, pointing to the top of the front arc on the south. The distance between any two zones is 15–20 km, and the entire spine is 70–80 km wide, showing a wedge shape wide in the north and narrowing in the south. After the spine tectonic traces pass through North Huaiyang to the north, they are connected to the N-S-trending 40-mile Changshan fold fault zone and iron ore zone on the Heyang Plain, crossing the E-W-trending tectonic zone. Geophysical prospecting data have revealed that the zone extends north of Yingshang and is actually the northern extension of the spine.[13] Based on the isotope age determination, the zone was generated at approximately 240 Ma in the early Indosinian. The N-S-trending compressional tectonic plane of the spine nearly disappears at Wangjiaba in the south and is replaced by N-W-trending and N-E-trending conjugate torsion faults or joints. It forms a checkerboard tectonic system, which is another manifestation of the spine and its disappearance.

The shield spreads in the Hong'an, Wangtianfan and Yuexi regions between the front arc and the spine and is a horseshoe-shaped section protruding to the south. The western and eastern sections are the Guankou arc-shaped tectonic zone and the Yuexi arc-shaped tectonic zone, respectively. The central section is composed mainly of Jinning and Yanshanian granite masses, with Proterozoic metamorphic rocks and fragments of the N-W-W-trending tectonic zone.

(c) *Western wing reflection arc*: This arc spreads in the Xiangyang–Fangxian–Wuxi region and forms an arc-shaped tectonic zone protruding slightly to the north, also known as the Fangxian arc. The arc top is located east of Fangxian and is composed of N-E-E-trending, N-W-W-trending, nearly E-W-trending or arc-shaped Proterozoic, Paleozoic and Triassic folds and thrusts. The northern section is dominated by the fold zone between the Qingfeng faults and the Jiudao–Yangkou faults, with strong compression and closed folds, and is reversed to the south. The faults are dominated by thrust faults protruding to the south, forming an imbricated shape. The southern section is dominated by the Shennongjia compound anticline and the Julongshan compound syncline, with relatively wide folds. The Shangnan–Danjiangkou NW-NWW-trending tectonic zone north of the Fangxian arc may also be the outer section of the western wing reflection arc. Its main body is the Huangling anticline composed of nearly N-S-trending short-axis rock blocks. The core is composed of the Paleoproterozoic metamorphic rock system and Jinningian granite mass, and the two wings are surrounded by Paleozoic and Triassic systems, with obvious unconformities. The Huangling anticline and the Fangxian arc are arranged in an arc shape and can be regarded as the spine of the reflection arc. The reflecting arc shields on the western wing are the Zigui Basin and the Jingdang Basin located on the western and eastern sides of the Huangling anticline, and they are coal-bearing basins composed of the Upper Triassic and Jurassic system, with gentle rock formations, weak deformation and well-preserved coal-bearing masses.

(d) *Eastern wing reflection arc*: This arc spreads in the Nanjing, Yangzhou, Zhenjiang and Changzhou regions and forms an arc-shaped tectonic zone protruding to the north, commonly known as the Ningzhen arc. The arc top is located west of Zhenjiang and is composed of NEE, NWW and nearly E-W-trending or arc-shaped folds and thrusts. The northern section is the Yangzhou compound anticline composed mainly of the Jurassic system.[14] The southern section is the Nanjing Faulted Basin composed mainly of the Triassic and Jurassic systems as well as Yanshanian granite intrusions. The middle section is the Yizheng depression, with large Cretaceous–Paleogene–Neogene deposits associated with developed arc-shaped faults with thrusting features, which is divided into an eastern section twisted to the right and a western section twisted to the left, corresponding to the stress field of the eastern wing reflection arc. The Jiashan–Jianhu arc-shaped uplift zone north of the Ningzhen arc may also be the outer section of the eastern wing reflection arc. The eastern wing reflecting the arc spine has always been regarded as the Maoshan tectonic zone. In recent years, some scholars have argued that the Maoshan tectonic zone was composed of thrusts formed in the Late Mesozoic, and some compressional tectonic planes intersected through the Ningzhen arc, which were the products after the Huaiyang epsilon-shaped tectonic system was finalized and could not be regarded as the spine of the Ningzhen arc. However, some scholars believe that there are many near-N-S-trending compressional tectonic planes in the Maoshan tectonic zone, with the western section slightly eastward and the eastern section slightly westward, showing an overall wedge shape that is narrow to the north and wide to the south. It extends to the top of the Ningzhen arc but does not intersect it. Its age is similar to that of the Ningzhen arc and can still be regarded as the spine of the Ningzhen arc. Based on lithofacies and paleogeography data, some scholars believe that there was a long-term depression basin in the regions of Changzhou, Lishui and Liyang, at least from the Triassic, on the inner side of the Ningzhen arc, and carbonate rocks and gypsum salt-bearing evaporites with thicknesses of more than 1,000 m were deposited in the early-middle Triassic. This area forms the oval low-lying region in the Shallow Sea–Gulf Lagoon depression zone. It was a relatively uplifted zone in the Late Triassic, and a volcanic rock basin was formed in the Liyong–Liyang–Jurong region in the Late Jurassic and Early Cretaceous and was deposited with thicker volcanic rocks.

Therefore, it is believed that there is a vortex on the inner side of the Ningzhen arc.

(2) Genetic Age of Huaiyang Epsilon-Shaped Tectonic System

The Huaiyang epsilon-shaped tectonic system has undergone a long growth and development process, which can be divided into an infant stage, a forming stage and a finalized stage.

There was an arc-shaped depression zone in the middle and lower reaches of the Yangtze River from the middle Carboniferous to the Late Permian. Until the middle Carboniferous, the southern deposition center was in the Tianmen and Daye regions of the western wing of the front arc and was N-W-trending and then passed through Yangxin and Jiujiang, and the eastern wing of the front arc was N-E-trending. The northern section spreads in Huangshi, Susong, Anqing, Chaohu and Nanjing, forming an arc-shaped settlement trough, and there are Susong, Anqing, Chaohu N-E-trending and Nanjing E-W-trending deposition centers along the zone. In the Late Permian, the deposition center was N-W-trending in the Liufang–Xialu region on the western wing of the front arc, turned to E-W-trending in the east, extended to the Anqing–Fanchang region of the eastern wing of the front arc, and then turned to N-E-trending. Therefore, from the formation and distribution in the middle Carboniferous–upper Permian, the middle and lower reaches of the Yangtze River were formed at the same age. In addition, for the formation age of the Huaiyang epsilon-shaped spine, based on the relationship between the N-S-trending thrusts and compressional fault zones in the Luotian–Yingshan region and the Yanshanian rock masses and dykes, the southern section of the spine formed mainly in the Late Jurassic. Based on the isotope age data of the Huoshou Iron Mine, the northern section of the spine might have been generated in the early Indosinian but finalized in the middle Yanshanian.[15] The growth and development of the spine were approximately synchronized with or slightly later than those of the arc-shaped tectonic zone.

(3) Ore-Controlling Effects of the Huaiyang Epsilon-Shaped Tectonic System

The two compressional tectonic zones, the front arc and the spine of the Huaiyang Epsilon-shaped tectonic system, have distinct ore-controlling effects on the iron-copper ore zone in the middle and lower reaches of the Yangtze River. Due to differences in lithofacies, there are obvious differences in the occurrence of iron, copper, sulfur, gold, gypsum, coal and other deposits in the inner, middle and outer arc-shaped tectonic lithofacies zones of the Huaiyang epsilon-shaped front arc. The magmatic rock activities are relatively weak in the inner zone, the Upper Paleozoic and Triassic evaporite systems are poorly developed, and the minerals are mainly copper ores, followed by iron ores and lead-zinc ores. There is a copper ore zone dominated by porphyry copper deposits in the Liuhe, Damachang of Chuxian County, Lujiang, Shaxi and Lucheng regions. There are hydrothermal copper deposits with the same age in the Huangmei Xishui, Gucheng and Zaoyang regions. The middle zone is a strong compression and torsion tectonic zone, with well-developed and large-thickness Upper Paleozoic and Triassic evaporite systems, composed of medium-to-basic volcanic rocks, subvolcanic rocks, and medium-basic, intermediate-acid, and partial-basic intrusive rocks widely distributed, dominated by iron ores, and accompanied by copper, sulfur, gypsum and alunite ores. There is an iron ore zone in the Ezhou, Daye, Huangmen, Luzong, Ningwu and Ningzhen regions. Because the western section is covered by the Jianghan Basin and the section further west is uplifted and eroded, the ore features are not studied in depth. The magmatic activities are dominated by neutral, intermediate and acidic rocks in the outer zone, and the intrusive rocks and volcanic rocks are more developed than those in the inner zone, but the intensity is not as high as that in the middle zone. The alkalinity is not as high as that in the middle zone, the Upper Paleozoic and Triassic evaporite systems are also less developed than those in the middle zone, the minerals are mostly copper ores with some iron, sulfur, and gold ores, copper mines are scattered in the Yangxin, Jiujiang, Guichi, Tongling, Fanchang, Lishui and Jurong regions, with the main copper mines in the middle and lower reaches of the Yangtze River. Therefore, these primary tectonic iron-copper metallogenic zones, the secondary iron-copper metallogenic zones and ore fields and deposits all reflect the gradual ore-controlling effects of the arc-shaped tectonic zone.

The Huoshou iron ore zone is located in the northern section of the spine of the Huaiyang epsilon-shaped tectonic system. The deposits and ore masses formed mainly in the Neoarchean–Palaeoproterozoic iron-bearing metamorphic rock system, and they are iron ore deposits reformed from clastic deposits or volcanic deposits. The iron-bearing rock system spreads in an E-W-trending direction regionally, but the spine of the Huaiyang epsilon-shaped tectonic system crosses the composite zone, forming a N-S-trending (centrally protruding to the west) fold zone.

The N-S-trending iron mine zones are arranged in rows from east to west and are hidden under the Quaternary system. There are mixed petrification and granite fossils in the ore-bearing tectonic zone. Isotope measurements indicate that the age of the original rocks is 2446–2580 Ma, while that of the regional metamorphism is 1716–1757 Ma. However, the biotite in the hornblende-bearing biotite quartz magnetite of the Zhouji Formation has a potassium-argon age of 250 Ma, indicating the important transformation and enrichment of iron ores in the active region of the Late Hercynian–Early Indosinian N-S-trending tectonic zone. Although the iron-bearing rock system is widely distributed in the Huoqiu Group from east to west, it has not been transformed into the N-S-trending tectonic zones in this region, so the industrial iron ore deposits and secondary iron ore zones are not formed.[16] Therefore, the formation of the Huoshou iron ore zone is closely related to the compound transformation of the spine of the Huaiyang epsilon-shaped tectonic system.

7.1.3 *Yunnan Epsilon-Shaped Tectonic System*

The Yunnan epsilon-shaped tectonic system is located in central-eastern Yunnan at 100°00'E–105°00'E and 23°30'N–27°00'N, with a length of 450 km and a width of approximately 350 km. It is a large-scale epsilon-shaped tectonic system.

(1) Main Features of the Yunnan Epsilon-Shaped Tectonic System
The front arc of this epsilon-shaped tectonic system spreads in the Nanhua, Jinning, Tonghai, Luliang, Xinping, Shiping, Honghe and West Shandong regions and is composed of four E-W-trending arc-shaped tectonic zones protruding to the south, including the Jinning Arc, Tonghai Arc, Yilong Arc and Honghe Arc from north to south, with widths of 150–160 km. The Jinning arc spreads in the Mouding, Yimen, Jinning and Xundian regions with a width of 3–4 km from north to south. The arc top is located 8 km southwest of Jinning, developed in the Pre-Sinian, Sinian and Lower Paleozoic and is composed of N-W-trending, N-E-trending, and arc-shaped folds and thrust faults, including the Mouding fault and the Banqiao fault. The Tonghai arc spreads in the Chuxiong, Eshan, Tonghai, Lunan, Qujing and Xuanwei regions and developed in the Lower Paleozoic, Upper Paleozoic and Jurassic, with a width of 7–8 km from north to south. The arc top is located 10 km southwest of Tonghai

and is composed of arc-shaped folds, thrust faults, and oblique thrust faults. The tectonic lines are relatively sparse on the western wing of the front arc, with the N-W-trending Quxi fault as the main trunk. The Cretaceous–Paleogene Chuxiong, Yao'an and other N-W-trending basins spread to the Dali and Xiaguan regions and gradually turn into N-W-W-trending, E-W-trending and even S-W-trending basins, forming the western wing reflection arc protruding to the north, commonly known as the Dali arc. The tectonic lines are dense on the eastern wing of the front arc and composed of a series of N-E-trending folds and thrusts arranged in a slightly ξ-shaped pattern. They spread in the Xuanwei and Weining regions and gradually turn N-E-E-trending, E-W-trending and even S-E-trending, forming the eastern wing reflection arc protruding to the north, commonly known as the Weining arc. The Yilong arc spreads in the Nanhua, Shuangbai, Xinping, Shiping, Jianshui, Mile, Luliang and Fuyuan regions, developed in the Paleozoic and Jurassic, with a width of 10–12 km from north to south. The arc top is located on the eastern margin of Yilong Lake and at 15 km southeast of Shiping and is composed mainly of large arc-shaped anticlines and faults. The western wing of the front arc has been disturbed and destroyed by the Sichuan–Yunnan N-S-trending tectonic zone, which has a more obvious manifestation in Shuangbai and Nanhua, and is composed mainly of N-W-trending folds and thrusts developed in the Jurassic. The folds are often S-shaped. The eastern wing of the front arc is composed mainly of N-E-trending thrusts, oblique thrusts and folds, such as the N-E-trending Qujing–Niushoushan fault, Panxi–Luliang fault, Xiaolongtan–Mile fault, Fule great anticline, and Mile and Qujing Paleogene–Neogene basins. The eastern wing of the front arc spreads in the Fuyuan, Tuchang and Shicheng regions and gradually turns to N-E-E-trending and S-E-E-trending, forming the eastern wing reflection arc protruding to the north. The arc top is located west of Tucheng, commonly known as the Tucheng arc. The Honghe arc spreads in the Nanjian, Yuanjiang, Honghe, Kaiyuan, West Shandong and Luoping regions, developed in the Paleozoic–Triassic and Jurassic, with a width of approximately 10 km from north to south, and the arc top is located 21 km east of Honghe County and is composed of arc-shaped folds, thrusts and compressional fault zones. The western wing of the front arc extends along the Yuanjiang and Lishe Rivers in a N-W-trending direction and is composed mainly of N-W-trending folds and compression thrusts, and some sections are reconnected with the Qinghai–Tibet–Sichuan–Yunnan ξ-shaped structure system and Honghe fault zone of the

South China epsilon-shaped tectonic system. It is separated from the Honghe fault zone east of Nanjian and north of Ailao Mountain and extends in the N-W-trending direction along the Lishe River, forming the western wing reflection arc (the Nanjian arc) protruding to the north near Nanjian.[17] The folds, thrusts, schistal zones of the metamorphic rock system and the Bouguer gravity anomaly zone also spread in an arc shape south of Nanjian and north of Wuliang Mountain. The Lancang River also spreads in an arc shape; the eastern wing of the front arc spreads mostly in the N-S-trending direction along the Nanpan River, with the N-E-trending Nanpanjiang fault as the trunk, extending to Luoping and Xingyi, and gradually turning to N-E-E-trending, E-W-trending and even S-E-trending, forming the eastern wing reflection arc protruding to the north, commonly known as the Xingyi arc.

The spine of the Yunnan epsilon-shaped tectonic system spreads in the Huidong, Wuding, Kunming and Yuxi regions on the northern side of the front arc. It is a N-S-trending linear compression tectonic zone, extending to the bank of the Jinsha River and gradually disappearing to the north. It extends to the north of Tonghai and gradually weakens to the south, measuring approximately 120 km wide from east to west and approximately 260 km long from north to south. It developed in the Pre-Sinian, Sinian and Paleozoic, reconnected to the middle zone of the Sichuan–Yunnan N-S-trending tectonic system and is composed of N-S-trending folds and thrusts, including the Daheishan syncline, Lufeng syncline, Sayingpan syncline, Malutang fault, Yanglin–Songming fault in the northern section and the Kunming Xishan fault and Sheshan fault in the southern section.

(2) Genetic Age of Yunnan Epsilon-Shaped Tectonic System

The Yunnan Epsilon-shaped tectonic system is composed of the Proterozoic Yukunyang Group and Sinian Paleozoic and Mesozoic systems and controls the emplacement of Indosinian granites. The Red River arc and the Middle-Lower Triassic marine facies are far thicker than those in adjoining regions. On top of the Honghe arc, Late Triassic terrestrial rift basins are developed. Along the Honghe arc, the Yilong arc, the sea arc and the spine, some small Cretaceous–Paleogene–Neogene settlement basins are scattered. It is speculated that the Yunnan epsilon-shaped tectonic system formed initially before the Triassic and even in the Late Paleozoic and finalized basically in the Indosinian Movement in the Late Triassic.

7.1.4 *Guangxi Epsilon-Shaped Tectonic System*

The Guangxi epsilon-shaped tectonic system is located in the central and northern Guangxi Zhuang Autonomous Region and south of Guizhou and Hunan Provinces, at 106°00'E–113°00'E and 23°00'N–26°30'N, measuring approximately 680 km long from east to west and approximately 360 km wide from north to south. It is a large epsilon-shaped tectonic system (Fig. 7.4).

(1) Basic Features of the Guangxi Epsilon-Shaped Tectonic System
The front arc of the Guangxi epsilon-shaped tectonic system spreads in the Donglan, Duan, Litang, Xiangzhou and Lipu regions. It is an arc-shaped tectonic zone protruding to the south, commonly known as the Guangxi arc. The front arc top is located in the Litang, Gula and Gantang regions between Kunlun Pass and Zhenlong Mountain and is composed of a series of arc-shaped or E-W-trending closed folds and thrusts, i.e., Litang anticline, Gula fault and Gantang fault. These faults are all N-S-trending and are obliquely intersected by a series of N-N-W-trending tension-torsion faults and shifted in joints. The Cretaceous and Quaternary basins also spread in an arc shape. The Yanshanian Kunlung Pass granite mass is emplaced on the southern portion of the arc top, which is the result of the strong extension on the arc top. The western wing of the front arc spreads through Daming Mountain, Duyang Mountains and Fenghuang Mountain and is divided into two sections by the Yishan E-W-trending tectonic zone near Baoping. The southeastern section is composed of N-N-W trending folds, thrust faults and oblique thrust faults, and its outer arc is dominated by the Damingshan great anticline and accompanied by multiple thrust faults. The axis of the Damingshan great anticline is composed of the Lower Paleozoic metamorphic rock system, and the two wings are composed of the Upper Paleozoic or Triassic system. The inner arc is dominated by the Du'an–Shanglin thrust zone and accompanied by small anticlines and synclines, including the Yaonan fault, Mashan fault, Shanglin fault, Tanghong fault and Hongdu fault, and the northeastern section is S-W-trending, forming an imbricated tectonic, which is intersected by a series of N-E-trending tensional or tension-torsion faults. The northwestern section is composed of N-W-trending folds and thrusts, i.e., the Donglan fault, Duyangshan anticline, Changpo fault and Hechi fault, forming the N-W-trending Duyang Mountains.[18] The eastern wing of the front arc spreads along Zhenlong Mountain, Dayao Mountain, Mengshan

Mountain, Haiyang Mountain and Dupang Mountain and is composed of N-E-trending and N-N-E-trending folds and thrusts. The outer arc is dominated by the Zhenlong Mountain–Dayao Mountain anticline and accompanied by multiple thrust faults. The axis of the Dayao Mountain great anticline is composed of the Lower Paleozoic metamorphic rock system, the two wings are composed of the Devonian system, and the entire anticline is S-shaped. The inner arc is dominated by the Sipai–Wuxuan thrust fault and accompanied by small anticlines and synclines with reverse S-shaped transitions, including the Wuxuan fault, Sipai faults, Tongmu fault and Lipu fault. The northwestern section is S-W-trending, forming an imbricated tectonic, which is intersected by a series of N-W-trending and N-W-W-trending tensional or tension-torsion faults.

The spine of the Guangxi epsilon-shaped tectonic system spreads in Rongjiang, Congjiang, Rong'an, Luocheng, Liucheng and Datang to the north of the front arc and is a N-S-trending tectonic zone measuring approximately 100 km wide from east to west and approximately 250 km long from north to south. It is composed of a series of N-S-trending folds, and thrusts can be approximately divided into northern and southern sections by the Yishan–Liucheng line. The northern section is the main body of the spine, which was generated earlier and is composed mainly of the Jiuwan Dashan great anticline, Motianling great anticline, Yuanbao Mountain great anticline and Siding great anticline. The core of the anticline is composed of the Proterozoic Sibao Group metamorphic rock system and early Paleozoic granites, reconnected and compounded on the ancient N-S-trending tectonic zone. Compression fault zones often appear in groups, including the Rongjiang fault zone, Tianhe fault zone and Longsheng–Liuzhou fault zone. The spine of the southern section was generated later and may have developed from the early shield. The folds are gentle and open, with the stratum oblique angle at mostly 5°–35°, and they gradually transition to the west. Therefore, the shield is not obvious.

The western wing reflection arc spreads in the Pingtang, Huishui, Luodian and Ziyun regions and is blurred due to the intersection of the N-S-trending tectonic zone. Only some arc tectonics protruding to the north can be seen in the Luodian–Nandan region, commonly known as the Moyang arc. In the Huishui–Duyun region north of the Moyang arc, the high and low points of a series of N-S-trending folds are connected, forming a profile approximately parallel to the Moyang arc. These points are obvious reflections of the mutual strengthening or weakening of the reflection arc and the N-S-trending tectonic zone at the superposition of anticline and syncline. The eastern wing reflection arc spreads in Linglingdao County and Jiahe regions, which is an arc-shaped tectonic zone protruding to the north that is composed of arc-shaped folds and thrusts, including the Zijin Mountain–Yangming Mountain–Tashan Mountain anticline and Daoxian–Jiahe syncline. In the Nanlan Mountain, Jianghua and Lianxian regions, there is a Sinian and Cambrian N-S-trending block, which is the main body of the reflection arc.

(2) Formation and Evolution of Guangxi Epsilon-Shaped Tectonic System

The Guangxi epsilon-shaped tectonic system is composed of the Proterozoic–Cretaceous system. Based on the screening of the tectonic traces on and below the regional unconformable surface and the analysis of the equal-thickness map of the Late Paleozoic deposit rocks, the formation and development process is as follows.

The Guangxi epsilon-shaped tectonic system was generated and formed initially in the Guangxi movement at the end of the Silurian. The distribution features of the tectonic traces under the unconformable surface between the Silurian and Devonian systems (Fig. 7.4) show that the basic outline of the Guangxi epsilon-shaped tectonic system was complete and that the positions of the front arc, spine and reflection arc were approximately consistent with those after finalization. The reflection arc and main body of its western wing are covered by the Upper Paleozoic and Triassic systems, and it is difficult to understand the actual conditions. In addition, the analysis of the equal thickness map of the first deposit caprock, i.e., the Devonian Lianhuashan Formation rock after the Guangxi Movement (Caledonian Movement), shows that it controlled the general appearance of the Lianhuashan deposits. However, the reformation and control formation indicate that there was an arc-shaped zone with thinner deposits protruding to the north in the Luodian, Wangmo, Nandan and Hechi regions and a nearly circular zone with thicker deposits in the Leye and Fenghuang Mountain regions in the Late Paleozoic. The former shows that the western wing reflection arc is an arc-shaped uplift zone in the Guangxi period, and the latter shows that the western wing reflection arc mainstay is a nearly circular vortex depression region in the Guangxi period and that they are at the same latitudes as the eastern wing reflection arc and mainstay, corresponding to each other.

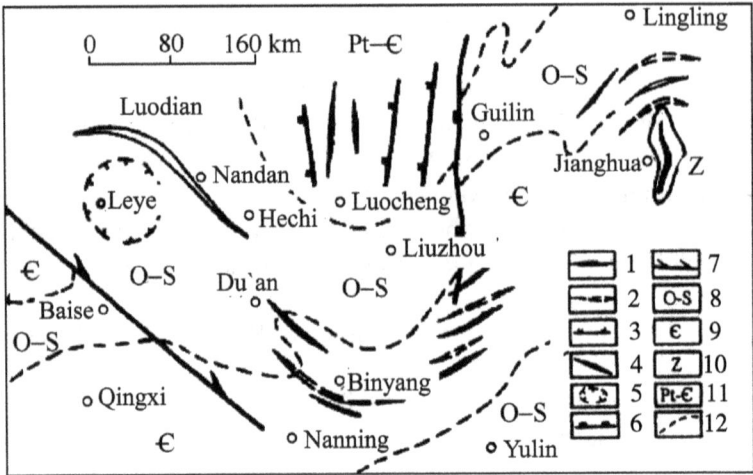

Fig. 7.4.   Schematic diagram of the Guangxi epsilon-shaped tectonic system.

Note: (1–5) Epsilon-shaped tectonic components (1) Anticline; (2) Syncline; (3) Fault; (4) Presumed uplift zone; (5) Presumed depression region); (6) N-S-trending fault; (7) N-W-trending fault; (8) Ordovician-Silurian system; (9) Cambrian system; (10) Sinian system; (11) Proterozoic-Cambrian system; (12) Presumed geological boundary.

Most of the Guangxi epsilon-shaped tectonic system was submerged by seawater from the Devonian to the middle Triassic and was further developed under the continuous action of the epsilon-shaped tectonic stress field. In this period, there was a vast shield and rapid depression, and the sediments had a thickness of nearly 10,000 m, dominated by carbonate rocks, which reflected that the area was a submarine depression area. The front arc and the northern spine subsided slowly, with the thickness of deposits from a few hundred meters to 3,000–4,000 m, dominated by siliceous rocks and argillaceous sandstones, reflecting that the region was a submarine uplift zone or uplift region. The western wing reflection arc is located in the Luodian, Wangmo and Nandan regions. The Devonian system is composed of Nandan-type deposits, namely, muddy, siliceous, and carbonate rock formations with rich floating organisms such as bamboo jade stones and ammonites. The mainstay of the western wing reflection arc is located in the Leye and Fengshan regions. The Devonian system is composed of Xiangzhou-type deposits, namely, carbonate formations with rich benthic organisms, i.e., brachiopods and corals.[19] In the

Yangmingshan region of the eastern wing reflection arc, the middle arc-shaped uplift zone protruding to the north was formed in the Devonian and controlled the accumulation of deposits in the strata.

The Indosinian Movement in the Late Triassic was the main finalization period of the Guangxi epsilon-shaped tectonic system (Fig. 7.5), with significant deformation and displacement in the Upper Paleozoic and Middle-Lower Triassic systems, forming fold fault zones of the front arc, spine and reflection arc. In the Gongcheng, Binzu and Duan regions, Jurassic and Cretaceous deposits are unconformable. At the same time, the Indosinian Movement also caused the spine to continuously expand to the south and the shield to gradually shrink to the center. The curvature of the front arc of the Guangxi epsilon-shaped tectonic system increased, and the angle between the two wings decreased. The spine advanced further southward, and the shield became narrower and even disappeared over time, indicating that the entire Guangxi region gradually shifted to the south during the development of the Guangxi epsilon-shaped tectonic system.

During the Yanshan Movement from the Jurassic to the Cretaceous, the Guangxi Shanzi was reactivated, and the Jurassic and Cretaceous folds and faults, as well as multiple strong intermediate-acid magmatic activities, complicated the Guangxi epsilon-shaped tectonic system (Fig. 7.5).

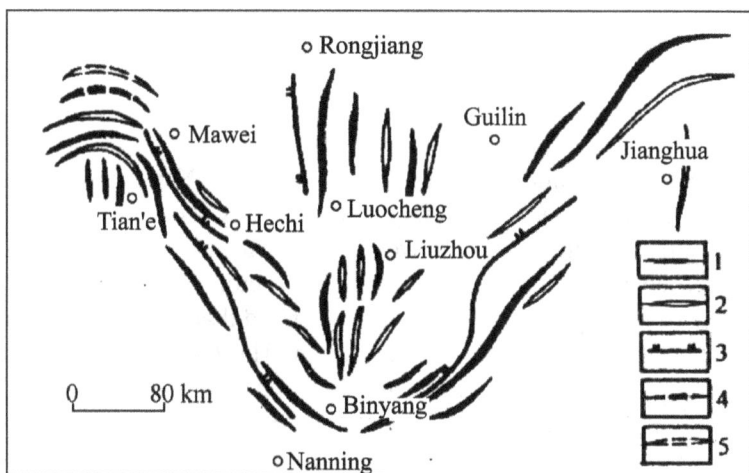

Fig. 7.5. Schematic diagram of the Guangxi epsilon-shaped tectonic system in the Indosinian.

Notes: (1) Anticline; (2) Syncline; (3) Fault; (4) Presumed anticline; (5) Presumed syncline.

(3) Ore-Controlling Effects of the Guangxi Epsilon-Shaped Tectonic System

The outer zone of the front arc of the Guangxi epsilon-shaped tectonic system is composed mainly of compound anticlines and thrusts, with strong compression and magmatic activity. The Guangxi and Yanshanian acidic and intermediate-acid intrusive rocks and lamprophyre vein groups are scattered on the surface or underground, controlling the formation of tungsten, tin, copper, lead and zinc deposits.[20] The inner zone of the front arc is composed mainly of synclines and thrusts, with strong compression, a small tectonic scale and weak magmatic activity, controlling the distribution of the Late Permian coal fields.

The northern section of the spine of the Guangxi epsilon-shaped tectonic system is strongly affected and reformed by ancient N-N-E-trending tectonics, with complex geological tectonics and frequent and intense magmatic activity. There are many tungsten, tin, copper, lead and zinc deposits at or near the axis of the Jiuwandashan Mountain, Motianling, Yuanbao Mountain and Siding compound anticlines, which are important Guangxi polymetallic deposits.

7.1.5 *South China Epsilon-Shaped Tectonic System*

In Zhejiang, Fujian, Guangdong, and southern Guangxi along the southern coast of Mainland China, a series of NE-NEE-trending Mesozoic tectonic magma zones and tectonic dynamic metamorphic zones are widely developed, while in southeastern Tibet, western Yunnan, and southern Yunnan in southwestern China, a series of NNW-NWW-trending Mesozoic–Cenozoic tectonic magmatic activity zones and tectonic dynamic metamorphic zones are well known. However, in the past, they were regarded as two independent geological tectonic units. Since the late 1970s, however, some geoscientists have noticed that there are only some possible genetic relationships between them. The Asian Geological Map Compilation Group, Liu Bo and Zhao Jian wei seperately proposed that they are the two wings of the front arc of an epsilon-shaped tectonic system called the South China Land Margin epsilon-shaped tectonic zone in 1979 and 1987, or the South China epsilon-shaped tectonic zone, and indicated the spread range, basic features, formation and evolution of this tectonic zone. Based on previous works, we carried out a comprehensive analysis of the latest geological results in the region and confirmed that there is a large epsilon-shaped tectonic zone in South China, spreading

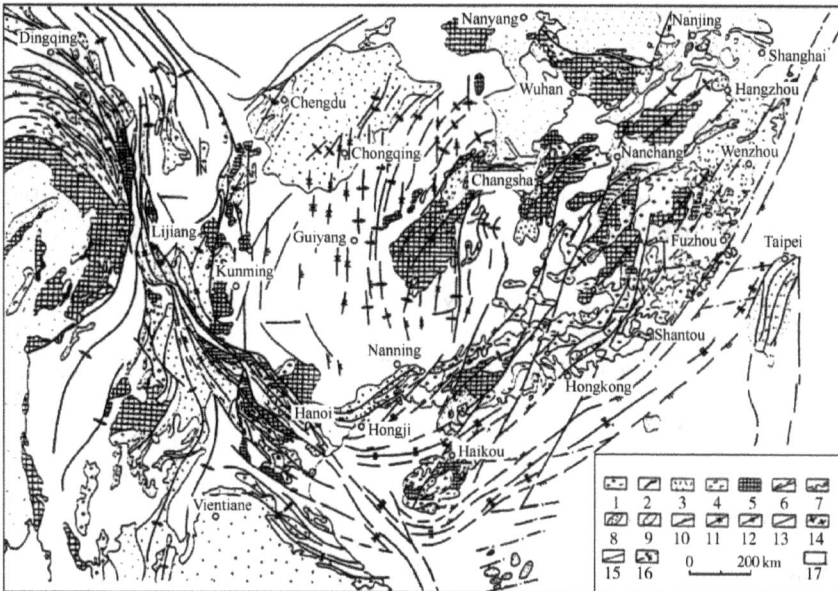

Fig. 7.6.   Schematic diagram of the epsilon-shaped tectonic system in South China (Zhao Jianwei, Wang Yeshun *et al.*).

Notes: (1) Intermediate-acid intrusive rock; (2) Ultrabasic intrusive rock; (3) Yanshanian intermediate-acid extruded rock; (4) Himalayan basic extrudates; (5) Luliang-Caledonian metamorphic rock block; (6) Tectonic dynamic metamorphic rock zone; (7) Schistal zone; (8) Yanshanian-Early Himalayan trough (basin); (9) Late Cenozoic trough; (10) Compression and compression-torsion fault zone; (11) Syncline; (12) Anticline; (13) Other system fault zones; (14) Other system compound anticline and syncline zones; (15) Concealed fault; (16) Concealed depression zone; (17) Paleozoic and Triassic distribution region.

throughout many provinces in South, Southwest and Southeast China, so it could be called the South China epsilon-shaped tectonic system (Fig. 7.6).

(a) *Basic outline of the South China epsilon-shaped tectonic system*: The main body spreads in the vast region south of the Qinling Mountains, the Nyainqentanglha Mountains to the west, the China Sea, the South China coast and the Taiwan Strait to the east, the north and middle of Vietnam, and the Xisha and Dongsha Islands to the south, spanning an area of 3×10^6 km².

(b) *Western wing and the reflection arc of the front arc*: The main body is developed in Beshula Range, Nushan Mountain, Yunling Mountain,

Gaoligong Mountain in the Nu River and Lancang River basins and Honghe River Basin, Wuliang Mountain and its southern region, Vietnam Shisongzaozai Mountain and Laos Fukote Mountain and the northern part of Changshan Mountain on the southeast. It is composed of a series of Mesozoic NNW-NW-trending fold fault zones, magmatic rock zones, and tectonic dynamic metamorphic zones and is often mitered or reconnected with the Sanjiang N-S-trending tectonic system and the Qinghai–Tibet–Sichuan–Yunnan reverse S-shaped tectonic system and scattered in the Sanjiang N-S-trending zone. However, after crossing the N-S-trending zone and the main zone of the reverse S-shaped tectonic zone, its distinct features become clear. The northwestern section is connected with the Nyainqentanglha arc zone, forming a deformation metamorphic zone with the Nyainqentanglha–Beshulaling compound syncline as the main body, and the intermediate-acid magmatic rock zone forms the main body of the western wing reflection arc. The reflection arc can be divided into northern, central and southern zones. In the northern zone, the Namco–Zhongba fault zone, Dulongjiang fault zone, and Bianba–Luolong–Basu fault zone are the southern and northern boundary faults, among which the Yanshanian intermediate-acid magmatic rock zone is intrusive in the form of a long and narrow zone and turns with the arc-shaped fold zones. The middle zone is the Dangxiong–Bomi fault uplift zone, and its axis is covered by the Late Yanshanian–Himalayan Linzhou–Gongbujiangda, Bomi–Zhuwagen, Zhongba–Baixue arc-shaped granites and the granodiorite zones, forming a central arc-shaped magma deformation metamorphic zone. The southern zone is composed of unidentified mixed rocks and deep metamorphic rocks in the Great Bend region of the Yarlung Zangbo River, with strong tectonic dynamic metamorphism and intermediate-acid intrusion activities, forming an arc-shaped metamorphic and mixed rock zone. The three tectonic magma deformation metamorphic zones are all arc-shaped, protruding to the north, and the main faults are twisted at opposite ends of the northern section and move at the same ends of the southern section. They pass through the Nujiang fault zone and extend to Weixi in the southeast intermittently, and the intermediate-acid magma intrusive activities are weakened.

The main fold zone on the western wing of the South China epsilon-shaped tectonic system is the Mesozoic Lanping–Luchun basin. The main fault uplift zones on the western wing are the Ailao Mountain–Diancang Mountain fault uplift zone and the Shisongzaozai Mountain fault uplift zone.[21]

(c) *Eastern wing of the front arc*: This region spreads on the southeastern coast and sea regions of China, starting from the South Yellow Sea on the north and passing Zhejiang, Fujian, Guangdong, southern Guangxi and Hainan Island, connecting with the arc top, and the tectonic zones, magmatic rock zones, and dynamic metamorphic zones are NNE-NE-trending.

(d) *Top of the front arc*: This region spreads mainly in southern Guangxi, Hainan and northern Vietnam and can be approximately divided into inner, middle and outer arc-shaped tectonic zones. From north to south, there are the Shiwanda Mountain–Hongji–Hanoi arc, Yunkaida Mountain–Bailongwei–Songzaozai Mountain uplift fault zone, Sangnu–Hejing–Hainan Island tectonic magma zone, and Fukote–Donghai–Hainan Shelf tectonically affected zone.

Inner zone: This zone is located between Shiwanda Mountain in Guangxi and Qinzhou Bay and passes through Hongji, the northern side of Hebei, Taiyuan and Zhanhua.

Middle zone: The main body is the Yunkaida Mountain–Bailongwei Island–Songzaozai Mountain fault uplift zone, and the exposed strata are mainly Lower Paleozoic and Neoproterozoic metamorphic rock systems.

Outer zone: This zone is discontinuous due to the intersection of N-W-trending faults, including the Laos Sangnu–Bilongshan fold fault zone, the Jurassic–Cretaceous volcanic rock zone, and the Yanshanian intrusive rock zone.

(e) *Spine*: On the inner side of the South China continental margin arc-shaped tectonic zone, a series of N-S-trending fold fault zones are developed. Among them, the N-S-trending tectonic zones at approximately 105°E–113°E end at the inner side of the arc top, do not cross the front arc, and mostly end near 30°N to the north.

7.2 Eurasian Epsilon-Shaped Tectonic System

The basic features of the Eurasian epsilon-shaped tectonic system were discussed by Pro. Li Siguang in "Introduction to Geomechanics".

It spreads east of the European continent and north and northwest of the Asian continent at 8°00'E–99°00'E and 35°00'N–70°00'N, with a length of approximately 5,000 km from east to west and a width of approximately 3,700 km from north to south. It is the largest known epsilon-shaped tectonic system in the Earth's crust (Fig. 7.7).

The front arc of the Eurasian epsilon-shaped tectonic system is an E-W-trending arc-shaped tectonic zone protruding to the south. The arc top is located in the Ashgabat and Mashhad regions on the southern side of the Karakum Desert, namely, the Kopit Mountain fold zone. The western wing of the front arc is N-W-trending, passes through the Caspian Sea and north of the Black Sea and is composed of the Alborz Mountains, the Caucasus Mountains and the Carpathian Mountains fold zones. Due to strong interference from other tectonic systems, the traces of the western wing reflection arc are unknown. The eastern wing of the front arc is N-E-trending, passing through the border regions of Kazakhstan and Afghanistan, reaching the mountains between the northwestern margin of

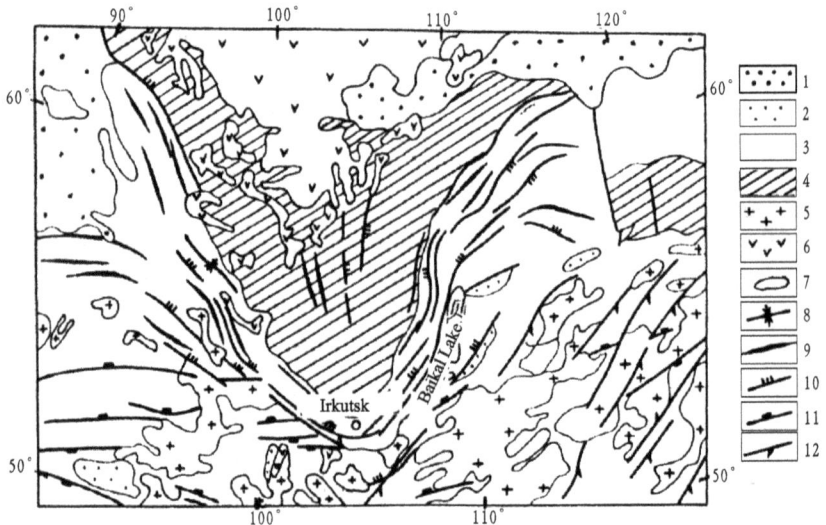

Fig. 7.7.   Schematic diagram of the Irkutsk epsilon-shaped tectonic system.

Notes: (1) Cenozoic basin; (2) Mesozoic and Cenozoic basin; (3) Paleozoic system; (4) Ancient block; (5) Intermediate-acid intrusive rock; (6) Basic rock; (7) Geological boundary; (8–10) Epsilon-shaped tectonic system composition (8) Compound syncline; (9) Fold axis; (10) Main fault; (11) Main fault in E-W-trending tectonic zone; (12) Nain fault in N-N-E-trending system.

China's Tarim Basin–southwestern Tianshan–northwestern margin of the Junggar Basin and southeastern margin of the Hasak Hills–Turan Plain. The northwest gradually turns E-W-trending and S-E-trending northwest of Junggar and forms the Altai Mountains arc-shaped fold zone protruding to the north along the Altai Mountains, namely, the western wing reflection arc, and then miter reconnects with the western wing and N-W-trending tectonic zones of the Mongolian arc. The spine of the Eurasian epsilon-shaped tectonic system is the Ural Mountains fold zone on the northern side of the front arc and is composed of a series of N-S-trending closed folds and thrusts, as well as N-S-trending ultrabasic and intermediate-acid rock zones. Its shield is composed of the Karakum Desert in the middle, the Russian platform on the west and the Siberian platform on the east. The section in China is described briefly as follows.

The eastern wing and its reflection arc of the Eurasian epsilon-shaped tectonic system spread in the mountainous region between the northwestern margin of the Tarim–Junggar block in China and the southeastern margin of the Kazakh Hills–Tulan Plain. It is a N-E-trending ξ-shaped tectonic zone spreading in an arc shape on the northern and northeastern margins of the Junggar Basin in the northeast, forming its eastern wing reflection arc. Due to the reversed connection and combination with the Tianshan–Yinshan E-W-trending zone and the Kunlun–Qinling Mountains E-W-trending zone, the eastern wing of the Eurasian epsilon-shaped tectonic system is discontinuous. Because the eastern wing of the reflection arc is rejoined and compounded with the Altai N-W-trending tectonic zone, it is strong, while the western wing is weak, but they incompletely surround the northern section of the Junggar Basin and are continuous and clear. Geophysical gravity and magnetic data have also indicated that there is a continuous arc-shaped anomaly zone. Judging from the formation data on the northwestern margin of the Tarim–Junggar block, the eastern wing of the epsilon-shaped tectonic system is composed of Hercynian subactive deposits and Ordovician Lower Paleozoic ophiolites and horny porphyries and siliceous mudstones, as well as pillow-shaped basalts and turbidite deposits along the Talbaha Platform–Wild Prass Slope and the Mayile–Naiming water arc tectonic zones. The middle-lower Silurian system is composed of volcanic and hard sand flysches and developed ophiolite melanges. In the Alai Mountain–Maidan–Kwakula (Kokshale Range) region northwest of the Tarim block, the Lower Paleozoic system is composed of upper Silurian variegated volcanic rocks. The eastern Alai Mountains contain Mesoproterozoic

Changcheng quartzites, phyllite sand and other shallow metamorphic rocks.[22] In the Upper Paleozoic system, the Alai Mountain–Kwakula Devonian system is composed of metamorphic clastic rocks and siliceous rocks sandwiched by basic volcanic rocks. The Carboniferous and Permian systems are composed of neritic carbonate rocks, terrigenous clastic rocks and flysches. The middle and upper sections of the lower Permian system are composed of continental intermediate acid volcanic rocks, tuffs and tuff sandstones. They are generally active one-time active deposits and were strongly deformed at the end of the Early Permian, forming N-E-trending linear tight folds, turning E-S-trending and forming an imbricated tectonic zone, and they were invaded by Late Hercynian granites. In its northeastern section and the northern margin of the Junggar Basin, namely, the Talbaha Platform–Huangcaopo and Mayile–Naiming water arc zones, the Devonian and Carboniferous systems are widely exposed, which are dominated by neritic clastic rocks and sandwiched by basic and intermediate acidic volcanic rocks and volcanic clastic rocks. The Permian system is exposed mainly on the margin of the intermountain depression zone and is composed of continental lake basin deposits and pyroclastic deposits. Their active development finally ended at the end of the Late Carboniferous, which was slightly earlier than that of the southwestern section, forming tight linear folds and an imbricated tectonic zone protruding into the Junggar Basin. The granites are relatively developed along this arc-shaped tectonic zone. The middle and Late Hercynian granites are intermittently exposed in the form of large rock foundations and rock branches and distributed in the zone along the axis of the compound anticline. Magma evolution history indicates the continuous evolution of quartz diorite-plagiogranite-granodiorite-biotite granites in the middle Hercynian, forming typical orogenic granites. Its alkalinity increased in the Late Hercynian and transformed into sodium-iron scoria granites, namely, postorogenic alkaline rocks. Geophysical prospecting data have indicated that this zone has a negative residual anomaly value of 0–30 mgL on the residual value map of the fifth trend surface of the Bouguer gravity anomaly, indicating relatively small crustal density and a certain thickness of silicon-aluminum shell base in this tectonic zone.

The main fault zones developed along with the N-E-trending and Junggar arc-shaped tectonic zones include the Wuqia fault and the Maidan River fault in the southwestern section, the Balluk fault, the Dalbut fault and the Ke-Wu fault in the northeastern section, and the Chaukar–Torangkuduk fault, the Narmand fault, and the Keketuo Sea–Karaschengarh fault on the eastern wing. Among them, the Wuqia fault is regarded as the hyperlithospheric fault, and it is a compression fault zone that separates the Tarim block from the active zone on the northwestern margin and a fault zone that has been active for a long time since the Paleozoic, which is composed of Late Paleozoic granites and ophiolites. The fault passes through the Korla deep fault and is inserted into the margin of the Tianshan Mountains E-W-trending zone on the northwest, which turns to S-W-W-trending and extends to the Alai Mountain region in the southwest. The Ke-Wu fault on the western margin of Junggar is also a hyperlithospheric fault distributed along the western margin of the Junggar Basin. It is a concealed fault zone and has been confirmed by geophysical prospecting and drilling surveys. It is a boundary fault zone separating the stable basin region and the active mountainous region. The Dalbut fault and Balerek fault parallel to this fault zone are both high-angle compression-torsion thrust faults, with a tectonic dynamic metamorphism zone width of 4 km, and there are ophiolites and ophiolite mélanges distributed in fragments along the fault zone, which formed mainly in the Late Hercynian and pushed into the basin in the Mesozoic, forming an imbricated tectonic zone. The tectonic zone was still active in the Cenozoic, and approximately five earthquakes occurred along this zone. Notably, the Ke-Wu fault gradually turned E-W-trending east of Karamay and connected with the Chaukar-Torangkuduke fault or the Naermande lithospheric fault zone, forming the arc fault zone protruding to the north on the Junggar block and the boundary separating the Paleozoic stable zone from the northern active zone in the basin. This arc-shaped zone is clearly reflected in gravity and magnetic geophysical data, and it has not crossed the Irtysh deep fault zone to the northeast.

The above data indicate that the eastern wing of the Eurasian epsilon-shaped tectonic system formed mainly in the Hercynian and that the Junggar arc was the inner reflection arc of the eastern wing, which is consistent with its western extension and the western wing. Its spine reconnected to the Late Hercynian N-S-trending tectonic zone of the Caledonian N-S-trending tectonic zone.

7.3 Irkutsk Epsilon-Shaped Tectonic System in South Siberia, Russia

The Irkutsk epsilon-shaped tectonic system was determined by Pro. Li Siguang in 1929. It is located in the Irkutsk region in South

Siberia, Russia, at 90°00'E–116°00'E and 52°00'N–62°00'N, measuring approximately 1600 km long from east to west and approximately 1,000 km wide from north to south. It is a large epsilon-shaped tectonic system (Fig. 7.7).

The front arc of the Irkutsk epsilon-shaped tectonic system spreads in the Sayan Mountains and Baikal Lake regions, and it is a V-shaped E-W-trending tectonic zone protruding to the southeast, commonly known as the Sayan Mountains–Baikal arc. The top of the front arc is located in the Archaean region near Irkutsk and is composed of a series of E-W-trending arcs, folds and thrusts, that is, the Hamar Mountain fold zone; the southwestern corner of Baikal Lake also extends in an arc shape. The western wing of the front arc is the Archaean, Proterozoic, Paleozoic and Mesozoic system between Irkutsk and Krasnoyarsk and is composed of a series of N-W-trending closed folds and thrusts. The Baikalian (Lvliangian) granite mass and Caledonian granodiorite mass are also arranged in a N-W-trending direction, namely, the East Sayan fold zone. This zone extends to the vicinity of Krasnoyarsk and gradually turns N-W-W-trending, E-W-trending and even S-W-trending, forming the western wing reflection arc protruding to the north, that is, the West Sayan fold zone, and its arc top is located south of Krasnoyarsk, at approximately 55°00'N.[23]

The eastern wing of the front arc is the Archaean, Proterozoic and Paleozoic system between Irkutsk and Bodebo and is composed of a series of N-N-E-trending closed folds and thrusts. The Baikalian and Caledonian granite masses are also arranged in the N-N-E-trending direction, that is, the outer Baikal fold zone, and Lake Baikal also extends to the vicinity of Bodebo in the same direction and then gradually turns to N-E-E-trending, E-W-trending and even S-W-trending, forming the eastern wing reflection arc protruding to the north, namely, the Vidim Plateau fold zone. The arc top is located north of Bodaibo, at 62°00'N, which is 7° from the top of the western wing reflection arc to the north. The spine of the Irkutsk epsilon-shaped tectonic system spreads in the Bratsk, Ustikut and Zigalovo regions on the northern side of the front arc. There is a N-S-trending uplift with large folds (with lengths up to 800 km) and a wide range (with a long axis length up to 200 km), the same trending compression zones and thrust faults, as well as E-W-trending extensional faults and the N-W-trending and N-E-trending conjugate torsion faults on the common flat Lower Paleozoic basement. The upper reaches of the Oka

River, Angara River and Lena River are also N-S-trending. The midline of the spine is located not on the bisecting line of the angle between the two wings of the front arc but slightly eastward, and the front peak of the spine extends to the south and gradually disappears on the inner side of the top of the front arc.

The shield of the Irkutsk epsilon-shaped tectonic system spreads in the Paleozoic and Mesozoic exposure regions in Teshet, Cheremukho, Angarsk and Tarasovo between the front arc and the spine; it is a horseshoe-shaped region protruding to the south. The western section of the shield is large and features a series of N-W-trending domes and basins. The eastern section of the shield is narrow and features a series of N-N-E-trending short-axis folds.

The largest feature of the Irkutsk epsilon-shaped tectonic system is its asymmetry. As mentioned, the top of the front arc does not protrude to the south but is sharply V-shaped. The centerline of the spine is located not on the bisecting line of the angle between the two wings of the front arc but slightly eastward. The eastern wing angle of the front arc is small, the tectonic line is N-N-E-trending, the western wing angle of the front arc is larger, and the tectonic line is N-W-trending. The shield is wide to the west and narrow to the east, making the overall shape skewed to the west.

The Irkutsk epsilon-shaped tectonic system basically formed in the Baikal Movement at the end of the Proterozoic and intersected with each other in the Caledonian Movement in the Silurian, indicating that the main stress trajectory network shape occurred in the flat beam. The locations with a tight combination of slab beam and the underneath rock mass (namely, the so-called basement layer or deeper layer) are generally consistent with the basement in the region where the concave surface of the reflection arc is relatively stable.

The depth of the epsilon-shaped tectonic system is still uncertain. Generally, the smaller the epsilon-shaped tectonic system is, the smaller the thickness of the affected rock strata is, and vice versa. Small- and medium-sized tectonic systems have not been found to belong to this type. As far as the epsilon-shaped tectonic systems discovered are concerned, the smallest system is more than 30 km long from the end of one reflecting arc to the end of another reflecting arc and more than 20 km long from the top of the outermost front arc to the farthest spine point from the front arc. The scale of this type of tectonic zone is still uncertain.

7.4 Toli Epsilon-Shaped Tectonic System in Turkey

This tectonic system is located in the central and southern parts of the Asian continent and is a large arc-shaped tectonic system. Its eastern wing is composed of the northeastern section of the Hindu Kush Mountains and the Sulaiman and Gildar Mountains. The folds in this section are not normally NE-SW-trending because they are obviously strongly compressed in the E-W-trending direction, resulting in an unusual bag-like fold region between Sulaiman and Gildal. If the folds that form these mountains are connected, especially if the folds scattered on the western areas of them are connected, it can be seen easily that they are wholly NE-SW-trending and then turn to the west before the Mekland region, and there are E-W-trending mountains on the sea floor approximately 96 km from the south of the Mekland coast. Farther northwest, these arc-shaped top folds that were cut off and sank to the sea floor reappeared on the eastern coast of the Oman Gulf and the Persian Gulf and became the main body of the Zagros Mountains (namely, the Iranian Mountains). The northwestern extension of these mountains gradually turns to the west and inserts into the eastern section of Turkey; that is, the Kardistan region is connected with the so-called Iranian tectonic zone in Turkey to form a reflection arc. North of this arc-shaped fold zone, where the eastern wing has been compressed and the arc top has been destroyed, there is an N-S-trending uplifted mountainous region. It cooperates with the above arc-shaped fold zone, forming the spine of an epsilon-shaped tectonic system. However, judging from the topographical map, this spine is not composed simply of N-S-trending folds but partly also of a group of geese-shaped folds. This group of geese-shaped folds has shown a tendency to twist to the south in the eastern region east of this fold zone (i.e., Balochistan shield, Afghanistan) corresponding to the western region (i.e., Iranian shield). The Oligocene strata are all involved in this epsilon-shaped tectonic system. Whether this tectonic system was active continuously after the Oligocene cannot be determined (Fig. 7.8).

There is a typical example of an epsilon-shaped tectonic system in Turkey. Its front arc is the arc-shaped Toros Mountains, its spine is the fold mountain zone in central Anatria, and Ankara is located west of this fold uplift zone. This uplift fold zone contains many complex tectonic components, including (1) NE-SW-trending folds and (2) N-S-trending folds and other types of compression zones. Notably, this complex wedge-shaped fold uplift zone spreads and tends to become narrower from north

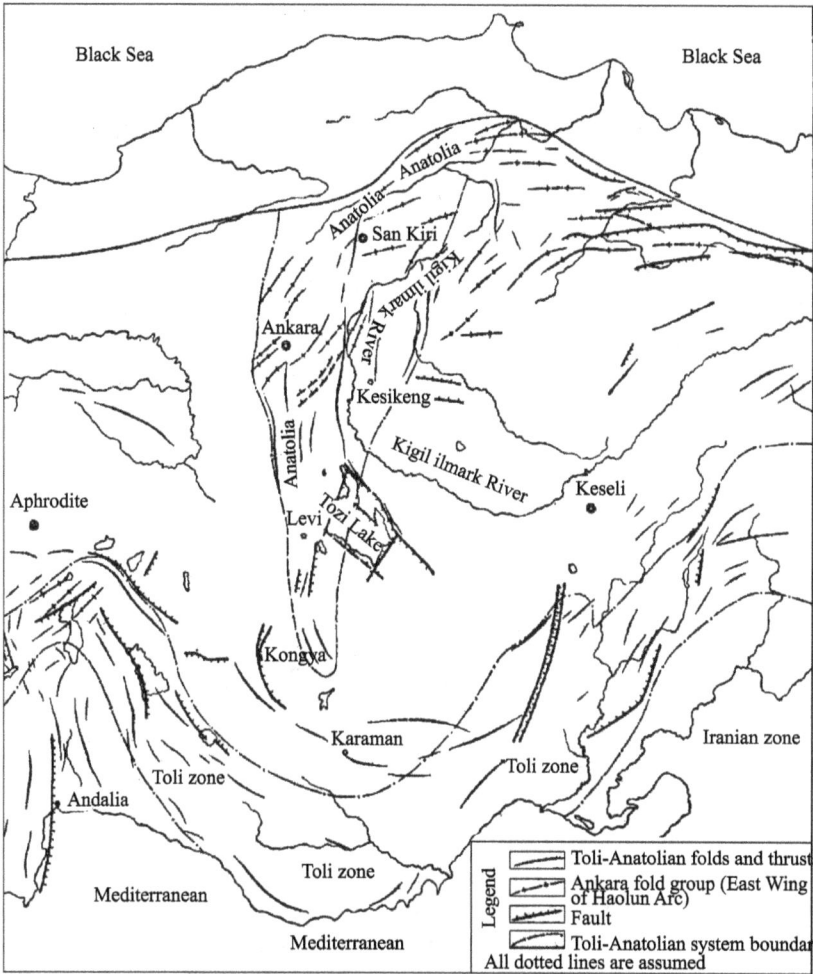

Fig. 7.8. Toli (Torros)–Anatorian epsilon-shaped tectonic system in Turkey (from the 1/800,000 Turkish geological tectonic map published by the Turkish Institute of Geology and Mineral Exploration. The scale of this map is approximately 1: 5500000).

to south and completely disappears at a considerable distance from the top of the front arc. The N-E-trending folds north of the spine turn to the east in the northeast, and they are likely part of the eastern wing of another epsilon-shaped tectonic system to the west of this epsilon-shaped tectonic system. The Toli–Anatoria epsilon-shaped tectonic system was formed at the beginning of the Tertiary period and was basically finalized in the Alpine Movement.

There may be the front arc of another epsilon-shaped tectonic system west of the above Toli–Anatorian epsilon-shaped tectonic system, which is generally known as the Heron arc. As mentioned, the easternmost section of its eastern wing is likely compounded with the northern section of the Toli–Anatoria mountain tectonic system. The southwestern section of its eastern wing may be located west of the horseshoe-shaped shield of the Toli–Anatoria mountain tectonic system. Farther southwest, it passes through the southern Rhodes, Kasos and Crete Islands, forming the top of the front arc. Its western wing passes through western Greece, forming Ionian and Pindus fold zones. Farther northwest, it extends to Albania, the southwestern coastal region of Yugoslavia and the southern Dalmacia Islands, showing a reflection arc shape.[24] Its spine should be the checkerboard tectonic zone in the Cyclades Islands and is most developed around the Naxos Islands. However, most of it is submerged under seawater, and only a N-S-trending compression zone appears on Sibos Island.

7.5 Gadom Epsilon-Shaped Tectonic System in France

This tectonic system is located in central and southern France based on the arc-shaped tectonics that existed in the Late Paleozoic and is collectively called the Gadom fold zone. Its eastern wing reflection arc is approximately parallel to the Jura Mountains. Its front part and the front sections of its two wings form the eastern, southern, and western arc-shaped fold zones on the Central Plateau. The latter section of its western wing is composed of Jurassic and Cretaceous rock masses, passes through the east of the Santongri region, gradually turns from N-W-trending to N-W-trending and enters the southern end of the Almoriga ancient fold block, forming a slight reflection arc. In the middle of the Central Plateau north of this arc top, there is a N-S-trending zone covered by the Early Tertiary rock mass, indicating that this N-S-trending depression zone (Overni Mountains) was compressed in the E-W-trending direction before the Tertiary. In the north, through the Loire River, there are N-S-trending ancient rock formations uplifted in the central basin in the Morfanregion. Whether this N-S-trending tectonic zone is also a part of the spine of the epsilon-shaped tectonic system in southern central France will be studied further.

7.6 Epsilon-Shaped Tectonic System in England

The epsilon-shaped tectonic system that appeared in central and northern England was mainly a product of the Hercynian Movement. Its western wing reflection arc is located in northern Wales and surrounds the Langollen spiral tectonic system. Its western wing is located between England and Wales, and its arc top and eastern wing are buried under the Jurassic and above rock masses. The results of paleogeographical research and geophysical prospecting indicate that its eastern wing bypasses the Oxfordshire and reaches the vicinity of Worthy Bay and that its spine forms the Pennin Mountains on the back spine of England. A part of the fold zone that forms the front arc of this epsilon-shaped tectonic system was damaged and partially buried under the new rocks, so it is not clear on the ground, but the boundary of the Triassic red beds deposited on the horseshoe basin between the front arc and the spine clearly shows the boundary of the concave surface of the front arc and the spine scale. The Central Plains of England are located between the back of the arc top and south of the spine. Although this epsilon-shaped tectonic system was basically completed in the Late Paleozoic, it was active continuously after the Jurassic and before the Cretaceous because the underground folds that form its eastern wing affected the Jurassic rock masses. The tectonic outline of the eastern wing of this epsilon-shaped tectonic system controls the distribution of the so-called coal fields in the United Kingdom.

7.7 Epsilon-Shaped Tectonic System in North America

This tectonic system is located in North America. It is a large-scale ancient epsilon-shaped tectonic system. The eastern wing of the front arc of this ancient epsilon-shaped tectonic system is consistent with the Appalachian Mountains and the forerunner of these mountain fold zones, namely, the Appalachian trough. It extends to Nova Scotia and Newfoundland to the southwest and passes through complex local twists, faults and new strata. Although the conditions are not fully clear, they generally turn from S-W-trending to E-W-trending (which may be compounded with the E-W-trending complex tectonic zone here) and to N-W-trending and then mix with the ancient Cordillera fold zone in the northwest. Behind this large arc-shaped tectonic zone protruding to the south, there is no ground syncline or strong fold zone, but there is a large small-amplitude dome zone, whose axis is N-S-trending with some local curves. A. Ketz has already pointed out the existence of this uplift zone, known as the spine of North America. All of them are symbols of an ancient and large epsilon-shaped tectonic system on the North American continent, which might have appeared in the Caledonian Movement and was finalized in the Helsinki Movement.

Strangely, the eastern wing of the front arc of the epsilon-shaped tectonic system suddenly disappears on the western coast of the Atlantic Ocean, while some fragments of the same Caledonian strong folds appear north of Ireland, northwest of Scotland and west of Scandinavia.

7.8 Cincinnati Epsilon-Shaped Tectonic System in Southeastern North America

In southeastern North America, there is an ancient epsilon-shaped tectonic system, which was finalized in the Late Carboniferous. Its spine is the so-called Cincinnati axis, and the eastern wing of its front arc is compounded with the Blue Ridge southwest of the Appalachian zone. In addition, a part of its western wing is covered by Late Carboniferous rock formations.

In western North America, there may also be an epsilon-shaped tectonic system compounded on the Cordillera N-S-trending large tectonic zone. It formed approximately in the Lebad period, namely, the period before the Ranami Movement. Its front arc spreads along the Coastal Mountains from the Olympia Dome in the north to south of Los Angeles in the south. The top of the front arc is located near Mendesino Point, and both ends show a reflection shape. Against the top of the front arc, east of the nearly N-S-trending fold zone near Great Salt Lake in Utah, an E-W-trending uplift zone composed of Precambrian strata suddenly appears, with a length of approximately 240 km. There are the Gelinhe Basin and Waxiaji Basin to the north and the Youyingta Basin to the south. The eastern extension is blocked by the controlled Nami uplift fold zone and the N-S-trending basin accompanying this uplift fold zone on the west. In a region with strong E-W-trending compression, a large and long anticline suddenly appears, and there is no obvious sign of E-W-trending compression. It is difficult to explain from different generation periods. Therefore, it is the spine of an epsilon-shaped tectonic system or a fragment of the E-W-trending complex tectonic zone.

In other locations in the Northern Hemisphere outside of China, there may be a number of epsilon-shaped tectonic systems, but they have not yet been fully determined. For example, in the Soviet Union, the basic outlines of the two large tectonic regions may be determined by some interconnected large and super-large fold zones and relatively stable blocks. One tectonic region is the Irkutsk Paddock and the ancient fold mountains around the paddock in the southern, eastern and western directions and the N-S-trending compression fault and large fault (compression) zone passing through the crystalline basement in the middle of the Paddock. The tectonic outline of this region was formed in ancient times. The other tectonic region spreads across the Eurasian continent, including the Urals and Russian platforms, and some strong fold zones along its southwestern margin, the Siberian platform, and the Central Asian mountains and some strong compression zones along its southeastern margin. The outline of the tectonic region formed at the end of the Paleozoic. However, because the generation and activity periods of the tectonic zones that form them have not been fully determined, they are treated only as an epsilon-shaped tectonic system (Fig. 7.9).

7.9 Brazil Epsilon-Shaped Tectonic System in South America

The Brazil epsilon-shaped tectonic system, or the Andes epsilon-shaped tectonic system, is a large epsilon-shaped tectonic system with a front arc protruding to the west. It is spread throughout nearly the entire region north of 25°N in South America. According to research by Ning Chongzhi and others, the front arc of the epsilon-shaped tectonic system is composed of single and compound fold zones and fault zones and spreads to the north of the Andes Mountains, including some mountains and the East Pacific deep trench. The northern wing reflection arc is intersected by the E-W-trending tectonic zone, and the southern wing reflection arc is clear and compounded with the Andean N-S-trending tectonic zone. The E-W-trending Cretaceous–Tertiary depression zone in the Amazon River and the E-W-trending fold zone east of the Madeira River are its spines. The Guyana and Brazil blocks on both sides are horseshoe-shaped shields of the epsilon-shaped tectonic system. Judging from the available data, the epsilon-shaped tectonic system was formed in the Cretaceous and Tertiary, and the deep trench on the western coast of South America

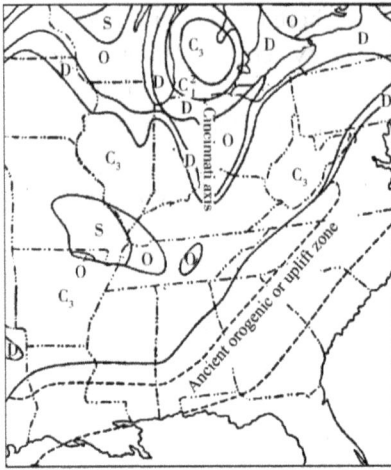

(a) Middle Pennsylvania period

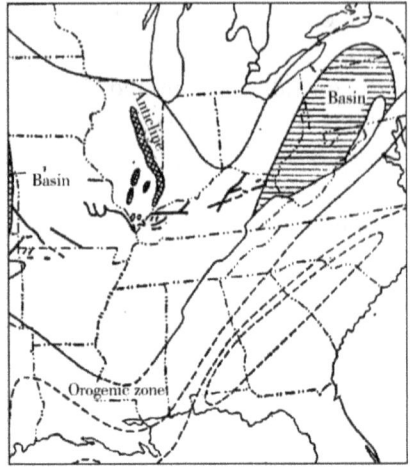

(b) Late Pennsylvania period

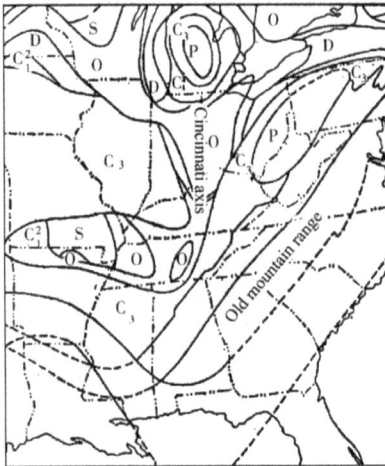

(c) Permian period

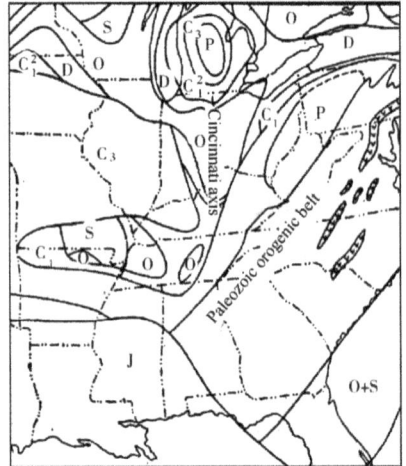

(d) Jurassic period

Fig. 7.9. Schematic diagram of the epsilon-shaped tectonic system that formed in the Carboniferous southeastern section of North America and its local transformation afterward (based on Eldrie's North American paleo-tectonic map).

Notes: O — Ordovician system; S — Silurian system; D — Devonian system; C_1^2 — Mississippi group; C_3 — Pennsylvanian group; C — Jurassic system; The anticline group with full intersecting lines shows the spine of the epsilon-shaped tectonic system formed in Late Pennsylvania; the thick lines and the orogenic zone show the front arc at that time; V represents igneous rock intrusions; and the chain lines represent state boundaries).

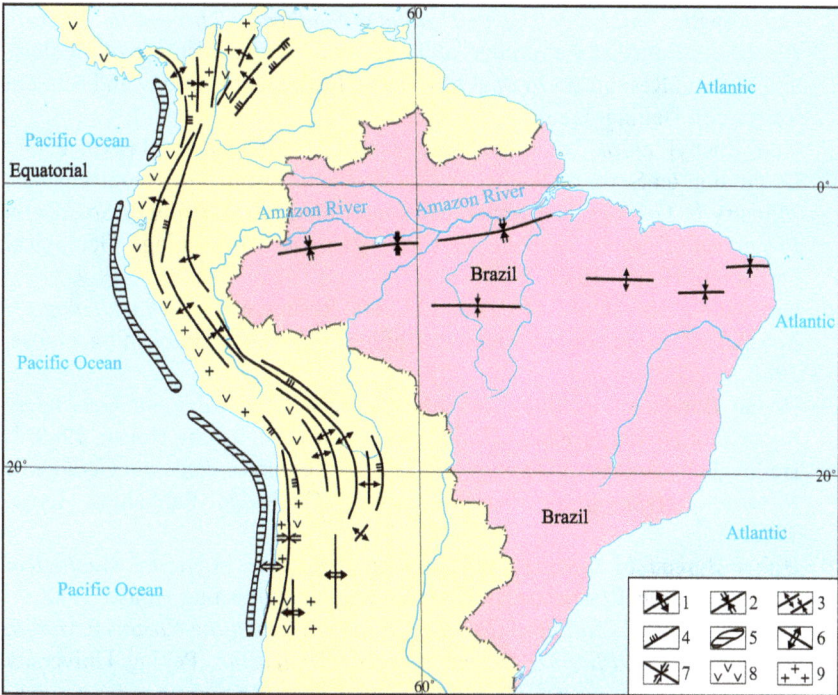

Fig. 7.10.   Simplified diagram of the 1 Brazil epsilon-shaped tectonic system.

Notes: (1–5) Compound anticlines, compound syncline, general anticline, compressional fault zone and deep trench of the Brazil epsilon-shaped tectonic system; (6, 7) Compound anticline and compound syncline of the Andes N-S-trending fault zone; (8) Volcanic rock; (9) Intermediate-acid intrusive rock zone.

indicated its obvious activities in the Neoid. Whether it existed in the pre-Cretaceous will be studied further (Fig. 7.10).

References

1. Fujian Bureau of Geology and Mineral Resources. *Regional Geological Records of Taiwan Province*. Beijing: Geological Publishing House, 1992.
2. Gansu Bureau of Geology and Mineral Resources. *Regional Geological Records of Gansu Province*. Beijing: Geological Publishing House, 1989.
3. Gao Zhenjia *et al*. Precambrian in the North of Xinjiang. *Precambrian Geology*, 1993, (6): 1–171.

4. Guo Yanghe. *Palaeogeographic Outline of Southeast Continent in the Early Paleozoic*. Editorial Committee of Journal of Nanjing Institute of Geology and Mineral Resources. Journal of Nanjing Institute of Geology and Mineral Resources. Beijing: Geological Publishing House, 1991.

5. Guo Zhenyi *et al.* Main Tectonic Systems in Shandong Province and Discussion on Several Geomechanical Problem. Institute of Geomechanics, Ministry of Geology and Mineral Resources (ed.). *Collection of Studies on Provincial Tectonic Systems in China (Volume 1)*. Beijing: Geological Publishing House, 1985.

6. Guangdong Bureau of Geology and Mineral Resources. *Regional Geological Records of Guangdong Province*. Beijing: Geological Publishing House, 1988.

7. Henan Bureau of Geology and Mineral Resources. *Regional Geological Records of Henan Province*. Beijing: Geological Publishing House, 1989.

8. Heilongjiang Bureau of Geology and Mineral Resources. *Regional Geological Records of Heilongjiang Province*. Beijing: Geological Publishing House, 1993.

9. Hubei Bureau of Geology and Mineral Resources. *Regional Geological Records of Hubei Province*. Beijing: Geological Publishing House, 1988.

10. Hu Xiao *et al. Evolution of Continental Margin along the Northern Margin of North China Platform during the Early Paleozoic*. Peking University Press, 1990.

11. Huang Hanchun. Xiyu Tectonic System in the West of China. *Editorial Board of Journal of Institute of Geology. Journal of Institute of Geology (#4)*. Beijing: Geological Publishing House, 1983.

12. Jilin Bureau of Geology and Mineral Resources. *Regional Geological Records of Jilin Province*. Beijing: Geological Publishing House, 1988.

13. Jiangxi Bureau of Geology and Mineral Resources. *Regional Geological Records of Jiangxi Province*. Beijing: Geological Publishing House, 1984.

14. Shao Yunhui. Yanshan Arc-Type Tectonic Belt and Its Derivatives Vortex Structure. *Institute of Geomechanics, Ministry of Geology. Collection of Geomechanics (#2)*. Beijing: Science Press, 1964.

15. Shui Tao. Geological Structural System of Zhejiang Province. *Institute of Geomechanics, Ministry of Geology and Mineral Resources. Collection of Studies on Provincial Tectonic Systems in China (Volume 1)*. Beijing: Geological Publishing House, 1985.

16. Sichuan Bureau of Geology and Mineral Resources. *Regional Geological Records of Sichuan Province*. Beijing: Geological Publishing House, 1991.

17. Boger, S. D., Wilson, C. J. L. Early Cambrian Crustal Shortening and a Clockwise P-T-t Path from the Southern Prince Charles Mountains, East Antarctica: Implications for the Formation of Gondvana. *Journal of Metamorphic Geology*, 2005, 23: 603–623.

18. Bonorino, G. G. Late Paleozoic Orogeny in the Northwestern Gondvana Continental Margin, Western Argentina and Chile. *Journal of South American Earth Sciences*, 1991, 4: 131–144.

19. Baruah, J. M., Handiquc, G. K., Rath, S. *et al.* Exploration for Paleocene-Lower Eocene Hydrocarbon Prospects in the Eastern Parts of Upper Assam Basin. *Indian Journal of Petroleum Geology*, 1992, 1(1): 117–129.

20. Basu, D. N., Banerjee, A., Tamhane, D. M. Geology of Bombay Offshore Basin, India. *Abstracts of the 26th International Geological Congress*, 1980, 26(1): 200.

21. Sun Dianqing, Gao Qinghua. *Geomechanics and Crustal Movement.* Beijing: Geological Publishing House, 1982.

22. Tan Zhongfu *et al.* Preliminary Study on the Neocathaysian System in the East of China and its Genetic Mechanism. *Acta Geologica Sinica*, 1983, 4(1) 45–52.

23. Tao Kuiyuan. Uniqueness of Volcanic Belt in the Southeast of China. *Volcanology and Mineral Resources*, 1991, 12(3): 1–14.

24. Ningxia Hui Autonomous Region Bureau of Geology and Mineral Resources. *Regional Geological Records of Ningxia Hui Autonomous Region.* Beijing: Geological Publishing House, 1990.

Chapter 8

S-Shaped or Reverse S-Shaped Tectonic Systems

Yuzhu Kang, Shuwen Xing, Zongxiu Wang, Zhihong Kang, Yinsheng Ma, Zhijiang Kang, and Zhihu Ling

Abstract

In this chapter, S-shaped or reverse S-shaped tectonic systems are introduced, including reverse S-shaped systems in Qinghai–Tibet–Burma, China, the western coast of North America and S-shaped systems in western South America and western Africa.

Keywords: Tectonic system; S-shaped; Reverse S-shaped; Type

The S-shaped tectonic system is one of the rotation-torsional tectonic systems. It was first identified and named by Pro. Li Siguang. Because its spatial distribution is ξ-shaped, this type of tectonic system is also called the ξ-shaped tectonic system or the S-shaped or reverse S-shaped tectonic system in this book.

This type of tectonic system is generally large in scale, its shape and composition are relatively complex, its compound forms with other tectonic systems are diverse, and mutual interference and utilization are also common. Based on its features, this tectonic system is generally divided into a head section, middle section and tail section, but they are connected to each other as a whole, without any boundary. Generally, its head section is composed of a set of strongly curved or even hook-shaped strong fold-fault zones. The middle section is composed of a number of strong parallel fold-fault zones that are generally nearly N-S-trending or NNW-SEE-trending, with some slightly curved sections protruding to the west or east. The tail section is also composed of strongly parallel fold-fault zones, showing a curved shape in the bending direction that is opposite to that of the head section. In this way, the head section, middle section and tail section together form a large reverse S-shaped tectonic system. The differences between this system and the general reverse S-shaped tectonic system are as follows. The former's head section generally shows a strong torsion phenomenon, and some fold-fault zones of the head section have a large curvature, while the tail section has a gentle curvature. The fold-fault zones around the head section may be scattered and discontinuous, and there may be several discontinuous, semicircular torsion tectonic zones with various curvatures around the head section.[1] The middle section is generally rejoined or mitered to the N-S-trending tectonic system. In most cases, the tail section is composed of a number of NW-SE-trending and nearly E-W-trending arc-shaped fold zones, and there is often a stable block in the center surrounded by these arc-shaped fold zones, which is opposite to the head section, forming a tectonic deposit depression or vortex.

8.1 Qinghai–Tibet–Burmese Reverse S-Shaped Tectonic System

The differences between this type of tectonic system and the ordinary reverse S-shaped tectonic system are as follows: (a) The former's head section generally shows a strong torsion phenomenon, and some fold-fault zones of the head section have a large curvature, while the tail section has a gentle curvature; (b) The fold-fault zones around the head section may be scattered and discontinuous, and there may be several discontinuous, semicircular torsion tectonic zones with various curvatures around the head section; (c) Its middle section is generally compounded with the N-S-trending fold zones; (d) The middle section is sometimes divided into two folds, sandwiched with a weak fold block; and (e) The tail section is often composed of a number of nearly E-W-trending arc-shaped tectonic zones.

The head folds of this type of tectonic fault often surround an uplift or depression block formed by horizontal torsion. There is often a depression zone submerged by seawater or against the sea on its two sides in the southwest.

A typical example of a reverse S-shaped tectonic system appears in western China and Southeast Asia facing the Indian Ocean. The head section of this reverse S-shaped tectonic system affects the Hengduan Mountains region in Qinghai, eastern Tibet, between Sichuan and Tibet, northwestern Yunnan, and northern Myanmar near India. The outer folds of the head section are scattered north of the Kunlun Mountains, including the Altun Mountains, the southwestern section of the Qilian Mountains near the Qaidam Basin, the Kukunuoer Ridge and the southeastern transition section of the Kunlun Mountains. The main components of the head section are distributed south of the Kunlun Mountains, including Hoh Xil, Bayan Har, Tanggula, Nyainqentanglha, the eastern section of the Gangdise Mountains, Bethula Ridge, Patkay and other mountains. These complex and large curved folds are largely folded and sometimes accompanied by large vertical faults and thrust faults. They have a similar arc shape and are obviously curved in the Qamdo and Yushu regions, especially in the Patkay region. From the northwest of Yunnan and the north of Myanmar to the south, the fold axis gradually turns to the south and becomes the middle section of this reverse S-shaped tectonic system, which is divided into eastern and western branches. The backbone of the eastern branch is the Ningjing Mountains between the Jinsha River and the Lancang River, including Shaluli Mountain, Daxue Mountain in the east, and Wuliang Mountain and Ailao Mountain in the south. This branch gradually spreads to northern Vietnam and Laos to the southeast and finally reaches the seaside. The western branch is the main branch, including Nu Mountain, Gaoligong Mountain and the N-S-trending mountains to the west, as well as the Arakan Mountains west of Myanmar. These N-S-trending mountains suffered strong E-W-trending compression in the Neoid, and there are also large faults parallel to the folds, with unclear characteristics. Farther south, a fold zone passes through southern Thailand and the Malay Peninsula.[2] Between this branch and the eastern branch, the Kolat Plateau is covered by Triassic and Jurassic rock masses.

The western branch spreads into the sea from the south of the Arakan Mountains and passes through Andaman and Nicobar Island and the islands of Sumatra and Java, forming the tail of this reverse S-shaped tectonic system. The tectonic section of these arc-shaped islands clearly reflects a strong side compression. On Sumatra and many islands off its southwestern coast, there are two uplift and large thrust fault zones formed by folds parallel to the coast. In the middle and southeast of the island, there are pre-Tertiary and Tertiary broom-shaped folds, with traces of torsion westward. There are also similar large folds and broom-shaped folds on the island of Java that are full of volcanic rocks.

The origin of the large reverse S-shaped fold zone spread in Qinghai, the adjacent regions of Qinghai–Tibet, western Yunnan, Burma, Sumatra and Java is still uncertain, but it reached the highest peak in the middle of the Tertiary period, namely, the Himalayan Movement or Alpine Movement. After the middle of the Tertiary, this orogenic movement did not stop completely (Fig. 8.1).

8.2 Reverse S-Shaped Tectonic System in China

8.2.1 *Qinghai–Tibet–Sichuan Reverse S-Shaped Tectonic System*

This system was called the Kang–Tibet reverse S-shaped tectonic system based on its spread in the past. It appears in southwestern China and passes through Southeast Asia and western Indonesia to the south or southeast. Its head section and peripheral fold-fault zones are scattered in Qinghai, Gansu, Tibet and the Northwest Sichuan Plateau. Its middle section spreads in eastern Tibet, western Sichuan–Yunnan and some regions of Burma, Thailand, Laos and Vietnam. Its tail section spreads mainly in Southeast Asia. The western branch passes through Andaman and Nicobar Islands in Indonesia and the western section of Sumatra and is reconnected and compounded with a torsion fold zone of the Oceania torsional tectonic system, i.e., the Sumatra and Java zone (Fig. 8.2).

The peripheral components of the head section of this reverse S-shaped tectonic system are scattered mainly in Qinghai, Gansu, Altun Mountains, the southwestern Qilian Mountains, South Mountain in Qinghai, and Solkuli, Qimantag, Jishi Mountain and other mountains and their corresponding tectonic zones north of Qimantag–Jishi Mountain (Animaqing Mountain). These complex and large curved fold-fault zones are largely folded and spread in very similar arc shapes. The outermost periphery of the head section may affect the areas north of the Qilian Mountains and the Hexi Corridor, and the northwestern margin may reach southeast of the Tarim Basin, while the northern margin is obviously blocked by Beishan Mountain.[3]

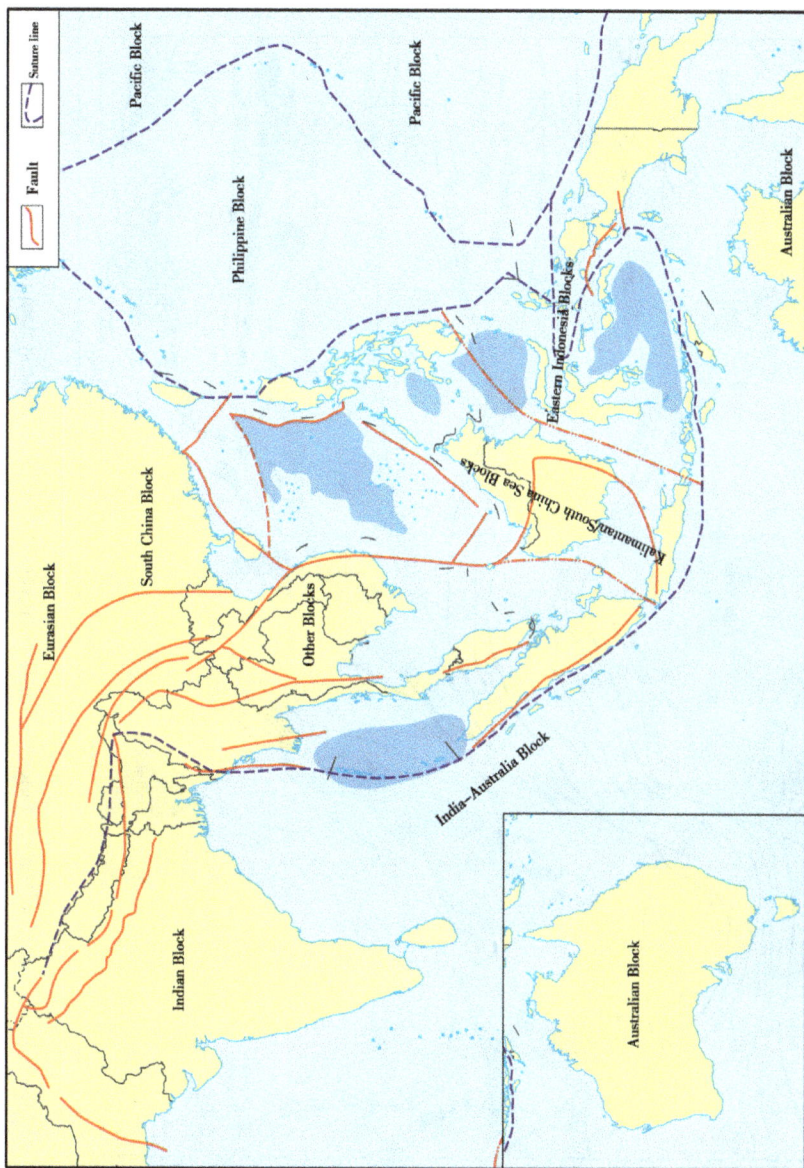

Fig. 8.1. Schematic diagram of the Qinghai–Tibet–Myanmar reverse S-shaped tectonic system.

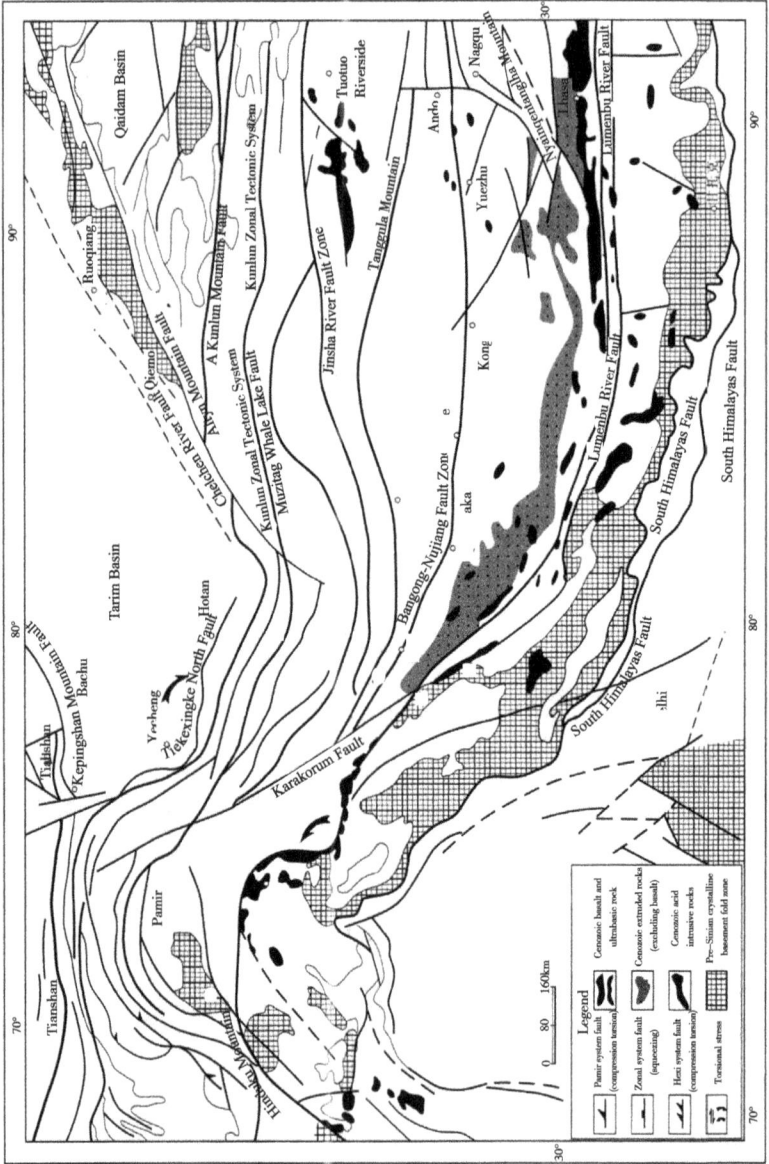

Fig. 8.2.   Schematic diagram of the Qinghai–Tibet reverse S-shaped tectonic system in southwestern China.

The Altyn tectonic zone is a combination of unique and regular tectonic traces formed by a series of large N-E-trending compression and compression torsional faults, linear folds, deposit troughs, strip intrusions and eruptive rocks. This tectonic zone is particularly obvious in the Altun Mountains and Beishan Mountain, with strong penetrating capacity and left-handed torsions. The formation and development of this tectonic zone are complex, and this zone has undergone multiple phases of activation. The main formation and activation periods are as follows: (a) Meso-New Proterozoic; (b) Paleozoic; (c) Hercynian; (d) Hercynian–Indosinian; (e) Jurassic; (f) Late Cretaceous; and (g) derived from the collision of the Indian block and the Eurasian block, only controlling the spread of the Cenozoic strata, formed in the Himalayan. The tectonic zone can be divided into three secondary tectonic zones: the Altyn uplift zone, Beiminfeng–Washixia fault depression zone and Cheerchen River fault uplift zone.

The tectonic features of the fault step zone southeast of the Tarim Basin are controlled by the Altyn fault system and the Cheerchen River fault system, with lengths of 1,000 km and widths of 40–140 km. The fault step zone can be further divided into the Cele uplift (IV_1), Minfeng Depression (IV_2), Qiwei Uplift (IV_3), Washixia Depression (IV_4) and Ruoqiangdong Uplift (IV_5) secondary tectonic units, with obvious left-handed torsional compression. In most of the zone, the Sinian–Triassic system is missing, and the Carboniferous–Permian and Jurassic systems remain in the depression, which is the highest uplift region in the Tarim Basin. The pre-Sinian system is exposed on the surface of the Niya No. 3 anticline and found at a drilling depth of 2,229 m in the Luobei #1 well, and it has a maximum burial depth of 5,000 m in the basement of the Minfeng Depression. The fault step zone is mainly the large-scale shear thrust nappe tectonic zone controlled by the Cheerchen River fault system. The main body of the fault step zone is composed of pre-Sinian metamorphic rocks and the overlying Cretaceous and Paleogene–Neogene systems (the Jurassic system is also confined in the Piedmont zone). The Danian thrust zone is overturned on the Guchengxu low uplift (Paleozoic). The evolution history of this fault step zone is unclear in the pre-Jurassic, and it is speculated that it was a part of the southeastern section of the tower and the Qaidam paleo-continent in the pre-Sinian. Its southwestern section was subjected to a transgression and developed a thin Carboniferous–Permian system with good oil-forming geological conditions in the Minfeng Depression in the late Paleozoic. Since the Mesozoic and Cenozoic, due to the control of the Altyn Mountains, this fault step zone has transformed into a piedmont depression, developed with the Jurassic coal-containing system and filled with Cretaceous and Paleogene–Neogene red molasses, with a total thickness of 4,000–5,000 m, forming part of the Tarim Basin.[4] In the late Himalayan cycle, the shearing and torsion movement was very intense.

8.2.2 S-Shaped Tectonic System in the Northwestern Jiaodong Peninsula

The S-shaped tectonic system in the northwestern part of the Jiaodong Peninsula is developed mainly along the margin and inside of the Linglong gneissic biotite granite and the early Yanshan Guojialing granodiorite masses. The Linglong granite mass itself has an S-shaped distribution, and the main surrounding rocks are the Archean–Paleoproterozoic Jiaodong Group, the Neoproterozoic Fenzi Mountain Group and the Penglai Group. Northeast of the granite mass, the two rock masses and the S-shaped fault are unconformable and covered by the Cretaceous system. This S-shaped tectonic system was formed at the latest at the end of the Jurassic. It is composed of granite and granodiorite masses, as well as four S-shaped main fault zones. They play an important role in controlling the formation and distribution of the gold deposit in the northwestern Jiaodong Peninsula. The following provides a brief introduction of this area from west to east.

(1) Sanshan Island Fault Zone

This fault zone is the westernmost cycle zone of the S-shaped tectonic system. A 2 km-wide mineralization alteration fault zone is exposed along the coast of the Sanshan Islands, with an overall fault width of 30–200 m, a trend of 40° north to east, and a southeastern dip angle of 35°–40°. It is developed on the side of the granites in the contact zone between the Linglong gneissic biotite granites and the Jiaodong Group and is a leftward compression and torsional fault. It is N-E-E-trending, extends to the Longkou region in the northeast and intersects the Tanlu fault in the southwest. Its southeastern section is the Quaternary S-shaped trough, forming the Tertiary syncline axis.

(2) Jiaojia Fault Zone
This fault zone is longer than 70 km and has a width of 80–200 m. Its middle section spreads at 25°–35° north to east, its southern section is developed in the Jiaodong Group and is N-E-trending, its northern section intersects the Linglong granites and the Shangzhuang mass, partially following the contact zone between the rock mass and the surrounding rocks, and gradually turns from N-E-trending to nearly E-W-trending, with an inclination angle of 30°–40° to the north, forming a naturally curved tectonic zone.[5] There is a well-developed and continuously distributed compression-torsional dynamic metamorphic zone along the main section that controls the formation and distribution of gold deposits in this region.

(3) Linglong–Zengjiawa S-Shaped Fault Zone
This zone has a length of 100 km and is inclined to the south or southeast, with an inclination angle of 31°–45°. The main body of the middle-southern section spreads along the contact zone of the rock mass, and the Jiaodong Group and inserts into the rock mass near Linglong; it is NE-NEE-trending. It is intersected by the N-N-E-trending fault zone and emplaced by the late Yanshan granodiorites, and its eastward extension is unknown. In Zengjiawa and south of it in the middle section of the fault zone, the main body is N-N-E-trending, the southern section is NEE-E-trending, it has an overall S-shape, its northern section is rightward compression-torsional, and its middle section is leftward compression-torsional. The features of the occurrence of gold-bearing quartz veins controlled by the fault zone itself and its derivative tectonics suggest that the mechanical properties of the fault formation period and the mineralization period are different.Compression-torsional deformation occurred first, followed by tension-torsional mineralization, with reverse movement directions.

(4) Aishan Fault Zone
This zone is the easternmost tectonic component of the S-shaped tectonic system and is E-W-trending. It spreads along the southern margin of the Guojialing mass to the Aishan region in the west. It is NE-NNE-trending and extends to the south, which is left-staggered by a N-N-E-trending fault zone, with a tracked length of 30 km.

From east to west, the spacing between the four arc-shaped fault zones is 14 km, 20 km and 24 km, respectively. The first three fault zones are inclined in opposite directions, forming thrust tectonic troughs with gentle inclination angles, and they are compression-torsion oblique thrust fault zones. They form a complete S-shaped tectonic system on the S-shaped rock mass, which may be the product of the composite effects of the N-E-trending and E-W-trending tectonic zones in the region.

8.2.3 *East Sichuan–North Guizhou S-Shaped Tectonic System*

The tectonic system spreads in the eastern Sichuan fold zone east of the Huayingshan Mountain fault zone, extends to the Qutangxia region on the southern side of the Qinling Mountains in the north and extends to the Dalou Mountain and Wumeng Mountain regions in northern Guizhou in the south. It is composed of a series of early Yanshanian S-shaped fold zones. The middle section is composed of partitioned fold groups. The western section is scattered, with NE-NNE-trending in the eastern Sichuan region, and gradually turns into a N-E-E-trending and nearly E-W-trending ξ-shaped fold zone. It extends north of the central Guizhou uplift to the southwest, which is composed of ξ-shaped fold groups that turn from N-E-trending to N-E-E-trending. The fold groups at both ends are naturally turned, and anticlines and synclines are alternately arranged and accompanied by some fault zones.[6] The fold groups are composed mainly of the Paleozoic, Triassic and Jurassic systems. In the later period, they were reverse-mitered or rejoined by the N-N-E-trending Xinhua fold zones and the Sichuan–Guizhou N-S-trending fold zones. Because they are mainly compounded in the form of folds, the tectonic pattern is relatively continuous and complex in the region.

The S-shaped tectonic system is the product of the eastern shift of the northern sections and the western shift of the southern sections of the Qinling and Nanling E-W-trending tectonic zones. Its formation period is consistent with that of the N-E-trending tectonic system in eastern China, and it was formed by restriction and late displacement of the E-W-trending tectonic zone in the formation period of the N-E-trending tectonic system. This type of tectonic system is manifested in the Shanxi continental platform in North China and the Daxinganling tectonic zone in Northeast China. They are formed by the barriers and accommodations of strong E-W-trending tectonic zones (they do not pass through the Qinling, Yinshan and other E-W-trending zones) during the development of the N-E-trending system (Fig. 8.3).

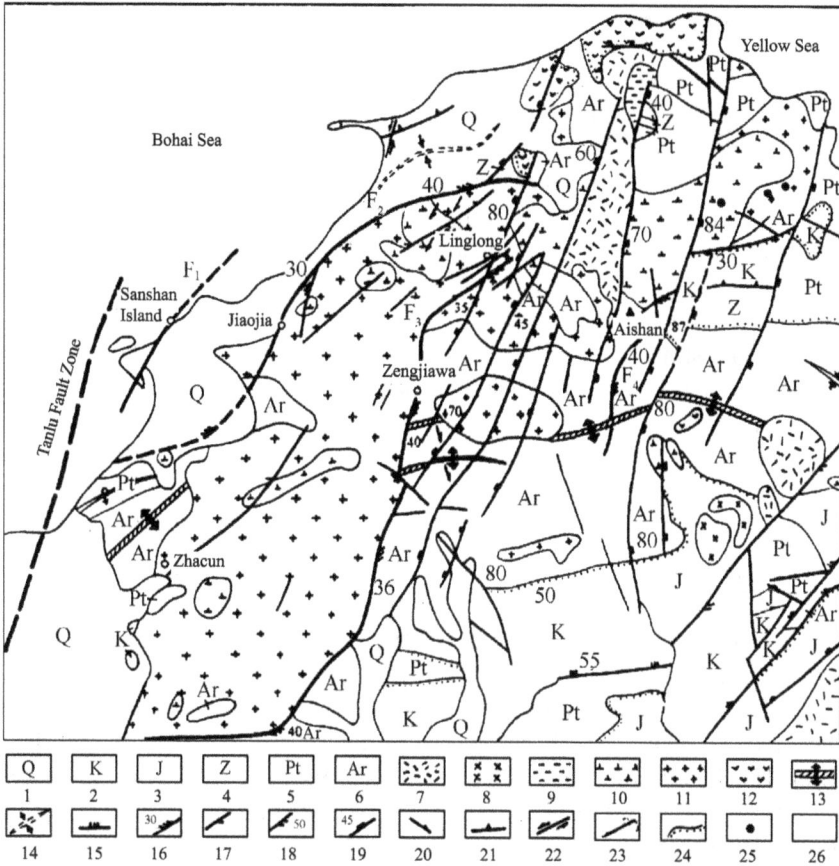

Fig. 8.3. Schematic diagram of the geological tectonic system to the east and west of the Jiaodong Peninsula.

Notes: (1) Quaternary system; (2) Cretaceous system; (3) Jurassic system; (4) Sinian Penglai Group; (5) Neoproterozoic Fenzishan Group; (6) Taiguyu-Paleoproterozoic Jiaodong Group; (7) Late Yanshanian granodiorite; (8) Late Yanshanian fine rock; (9) Late Yanshanian quartz diorite porphyry; (10) Early Yanshanian granodiorite (Guojialing granodiorite); (11) Late Indosinian gneissic biotite granite (Linglong granite); (12) Tertiary–Quaternary basalt; (13) Qixia compound anticline axis; (14) Tertiary syncline axis; (15) E-W-trending compression fault; (16) S-shaped tectonic and compression-torsional fault zone; (17) S-shaped tectonic and compression-torsional fault; (18) N-N-E-trending compression torsional fault; (19) N-N-E-trending compression-torsional fault zone; (20) N-N-E-trending tension-torsional faults; (21) Compression normal faults since the Tertiary; (22) Torsion faults since the Tertiary; (23) Unidentified faults; (24) Unconformably boundary; (25) Gold deposit; (26) S-shaped tectonic main fault number.

8.2.4 *Doranasay S-Shaped Tectonic System*

The S-shaped tectonic system is located in the Doranasay gold mining region in Habahe County, Xinjiang. This system has two mainstays that differ from other S-shaped tectonic systems. This twist tectonic was previously called the concatenated twist tectonic or the dual-core twist tectonic, but the most common combination of the cycle surface is S-shaped, reverse S-shaped or hyperbolic-shaped, as well as turbine-shaped, lotus-shaped and other types, so it is still classified as S-shaped and reverse S-shaped.

The cycle zone developed in the Devonian shallow metamorphic rock system. The strata, folds, faults and schistosity zones are distributed along the two rock masses in a reverse S-shape; that is, they form a dual-core torsion tectonic system in torsion movement in a counterclockwise direction with two equiaxed biotite plagiogranite plants in the northwest and southeast as the mainstays.[7]

8.2.5 *Xiaseling Reverse S-Shaped Tectonic System*

This tectonic system spreads in the Xiaseling region of Zhejiang Province and features a twist core at the waist of the reverse S-shaped tectonic system, which is different from other S-shaped and reverse S-shaped tectonic systems. Its cyclic zone is composed of a series of tension-torsional fissures. These tension-torsional cyclic surfaces are filled with tungsten veins, forming the Xiaseling tungsten ore field. On the surface, it is a broom-shaped tectonic system composed of two sections diverging apart, converging toward each other, and connecting with each other. Exploration has revealed that their fissures and veins are connected together in the lower part, forming a tension-torsion reverse S-shaped tectonic system, and there is a large concealed granite mass in a deep location in the waist section. Judging from the available data and the intrusion of granite masses, the formation of reverse S-shaped tectonics, and the filling of tungsten veins, there was likely a clockwise horizontal torsion movement with the rock mass as the central axis in this region. This kind of torsion action is related to the counterclockwise torsion of the large N-N-E-trending compression torsion faults on both sides.

8.2.6 *Enkou–Douli S-Shaped Tectonic System*

The above S-shaped and reverse S-shaped torsion tectonic systems are all products of horizontal and upward torsions around a vertical axis.

The Enkou–Douli S-shaped tectonic system is a vertical torsional system, which is also common on the surface. As Pro. Li Siguang describes in his discussion of petroleum geology, this torsion tectonic type is produced in torsion twists or torsion, which was first discovered by Sun Dianqing *et al.* in the Qaidam region and called the Bagyauru S-shaped tectonic system.

The Enkou–Douli S-shaped tectonic system spreads in central Hunan, with the Enkou–Doulishan compound syncline as its main body. It is N-E-trending, and both ends are twisted in a nearly E-W-trending direction, showing an S shape on the surface, but the folds at both ends are reversed in opposite directions, showing an obvious vertical torsion effect along the horizontal axis. The strata distributed in the region are composed mainly of Carboniferous, Permian and Mesozoic systems. The compound syncline is composed of four large-scale folds and faults, including the Enkou syncline, Renshoutang syncline, Huangtuling anticline and Qilinshan–Ganxi thrust syncline, as well as some compressional and compressional-tensional faults. It has a significant effect on the later formation of coal seams and a significant control effect on the distribution and quality changes of coal seams in the region. That is, the thin coal zones of the mineable coal seams of the Permian Longtan Coal Group are distributed in regions with obvious S-shaped tectonic systems with larger curvatures, especially the coal seams located in the reverse wings of anticlines and synclines, which are more severely damaged. In the S-shaped tectonic system, the location with the highest metamorphism in the Longtan coal system is also located in the region with the most obvious and largest curvature.[8]

In addition, this S-shaped torsion tectonic system can also be found in Etouchang of Yunnan, and the Yunnan Etouchang iron deposit itself is an S-shaped ore zone distributed along the horizontal S-shaped anticline axis.

8.3 Reverse S-Shaped Tectonic System on the Western Coast of North America

There is also a large S-shaped tectonic system in North America facing the Pacific Ocean. Its head section is composed of the Alaska Peninsula, the Aleutian Mountains, the Chiu Kutch Mountains, the Alaska Mountains, the St. Elias Mountains formed by strong folds, as well as the Endicott Mountains in northern Alaska and the Mackenzie Mountains in northern Canada. Its middle section is composed of the Cordillera

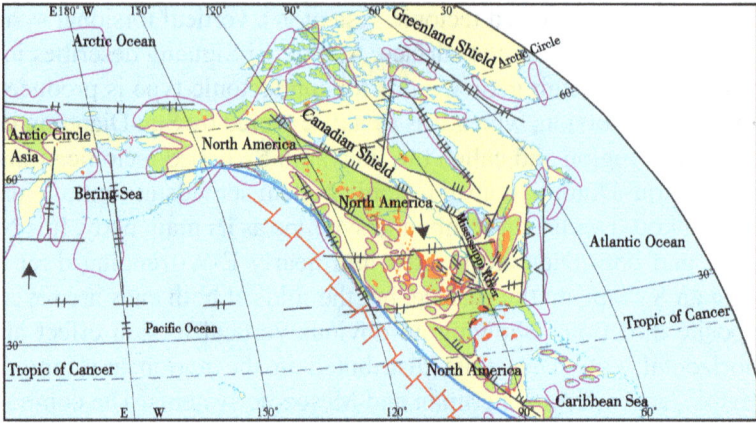

Fig. 8.4.   Schematic diagram of the reverse S-shaped tectonic system western coast of North America.

Mountains, which were transformed from the Cordillera syncline and extended to the Bashuma in the southeast. These large and complex mountains can be divided into two-fold zones: the Coastal Mountains to the west and the Rocky Mountains to the east. A long blocky zone is sandwiched between these two skirt zones. The two folds and the block between them extend 31°N to the south, and they are curved more significantly to the southeast, forming the E-W-trending Madre Mountains. These mountains should be the tail section of this large reverse S-shaped tectonic system. In the south, it is the South Madre Mountains, with an obvious tendency to bend to the east.[9] After passing through Guatemala and Honduras, the same fold zone spreads into the Caribbean Sea. The islands in Jamaica, Dominica and Puerto Rico belong to the tail section of this reverse S-shaped tectonic system, and they are approximately comparable to the island of Java in the tail section of the reverse S-shaped tectonic system facing the Indian Ocean (Fig. 8.4).

8.4 S-Shaped Tectonic System on the Western Coast of South America

8.4.1 *Tectonic Features*

Most regions of Peru, Bolivia, Chile and Argentina on the western coast of South America belong to the Andean orogenic zone. These regions can be divided into three zones from west to east: (a) *Front arc zone*: The slope zone and the far shore zone in the Pacific Ocean, with a large ocean region on the north, in which there are some basins, i.e., the Arauco, Tata and Valdivia basins in Chilean waters; (b) *Volcanic island arc*: The current mountain system, including the Cordillera mountain system and the lower mountains near the coast, in which there are still active island arc zones which are very wide and composed of late Mesozoic and Cenozoic island arc volcanic-deposit rocks, as well as Cenozoic continental coarse clastic deposits and many accretionary blocks; and (c) *Back arc zone*: The East Cordillera mountain system, etc. (Fig. 8.5).

8.4.2 *Evolution of Basins and Deposits*

The deposit basins in southern South America have a long and complex evolution history. Since they were activated in the basement in the early Paleozoic, their stratigraphy and tectonic features are complex, and they have experienced a long evolution history in confined sea regions, forming very good hydrocarbon sources and reservoir rocks with good oil-generating potential in several basins.

The evolution of the basins in this rea can be divided into two obvious stages: the pre-Andean stage and before the late Cretaceous. Due to the complex tectonic background, there are various basin types. After the Late Cretaceous, they are foreland basins mainly in the Cenozoic.[10]

From the Cambrian to the Late Jurassic, the basins and island arcs on the western coast of South America were nearly N-S-trending. The following N-S-trending stress was related to the breakage of the Gangwana mainland and the separation of the South Atlantic. There were multiple basin formation stages (Fig. 8.6).

(1) Cambrian–middle Devonian passive continental margin deposit basin
This basin is composed of Cambrian–middle Devonian terrigenous clastic rocks, carbonates and intrusive rocks along the western margin of the Brazil, Pune and Pampas shield regions.

(2) Carboniferous–middle Jurassic Intracontinental Rift Basin
The deposits in the Carboniferous–middle Jurassic intracontinental cracks are derived mainly from continental sources, and the marine clastic rocks are concentrated mainly in the foreland basin on the front margin of the

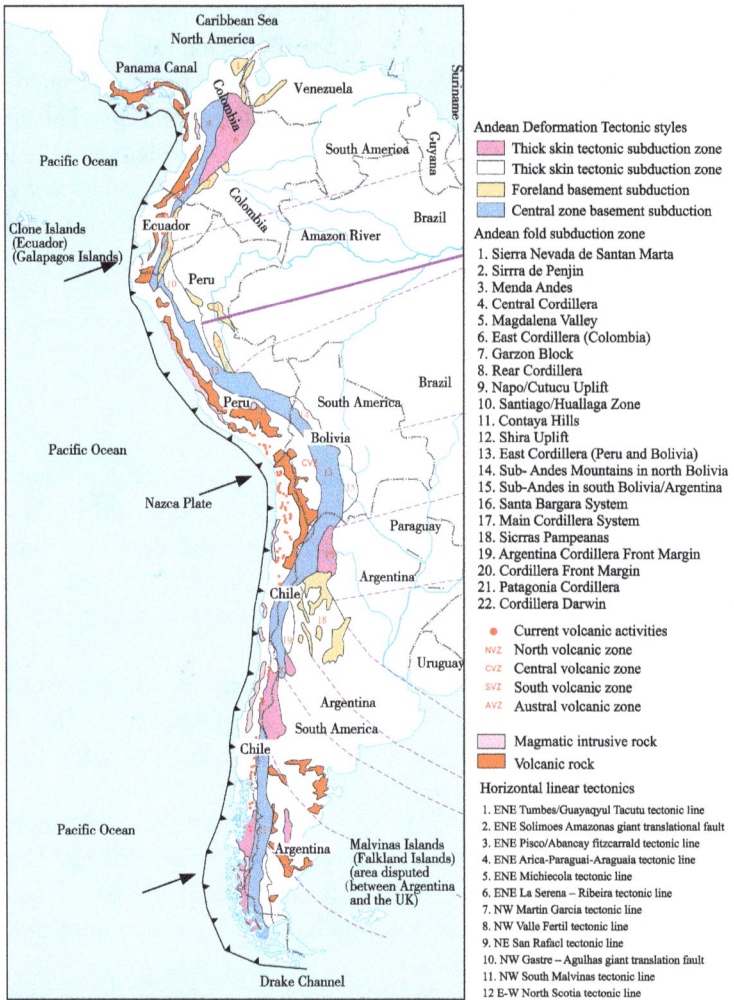

Fig. 8.5. Schematic diagram of the S-type tectonic system on the western coast of South America.

volcanic island arc west of the continent. This stage was active until the breakup of Gondwana and a large amount of magma intrusion in the Late Jurassic.

(3) Late Jurassic Rift basin
The extension in the Late Jurassic is the Andean foreland sequence marked by large-scale pan flooding deposits and closely connected between the

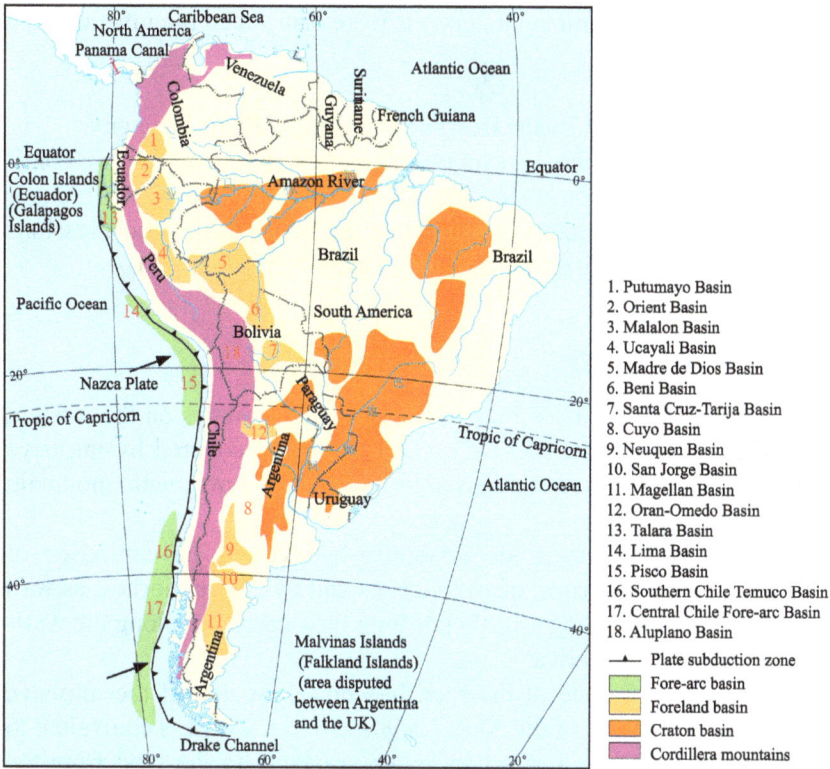

Fig. 8.6.   Distribution of major deposit basins western South America.

clastic rocks, evaporites and acidic volcanic rocks with the volcanic island arc system. The Patagonia acidic volcanic rocks cover the entire North Patagonia block and the Desaiado Craton, as well as some earlier basins.[11]

(4) Middle-Late Cretaceous Rift Basin
After crackup, a deposit column formed along the western margin of the Atlantic Ocean. The thick shales, limestones, evaporites and pyroclastic rocks are related to the middle and Late Cretaceous volcanic arc and back arc background. New long acid intrusive rocks and Andean rock masses formed in this period.

(5) Late Cretaceous–Paleogene Foreland Basin
The Late Cretaceous–Paleogene (Laramie) system was marked by the finalization of the Andean fold zone and thrust fault zone and the Andean rock mass. A curved foreland basin formed on the front margin of the

Andes, and passive continental deposits were distributed mainly along the eastern margin.

(6) Neogene Foreland Basin Destruction and Andean Orogeny
The Neogene marked the Andean orogenic movement and passive continental margin depression stage. The thrust faults provided a large number of sources for the foreland basin, and the small-scale transgressions spread over most of Patagonia and the Pampa Plain (Fig. 8.7).

8.4.3 *Intrusive Rocks*

A large number of intrusive rocks developed in the Andes on the western coast of South America, and 15% of the region is covered by intrusive rocks. Thus, the Andes Mountains are often named a magmatic mountain system.

The Andes Mountains are generally believed to be composed of granodiorites, daxodiorites, quartz diorites and pyroxene diorites, as well as fewer granites, forming a 1,200 km-long rock basement along the western coast of South America.

In the northern zone of the rock basement, the age of the intrusive rocks in the main body of the Andes is 90–120 Ma, which is equivalent to the middle Cretaceous, and decreases successively to the west from the Late Cretaceous (76 Ma) to the Eocene and Pliocene (15–25 Ma). Locally, there are later intrusive rock masses with ages of 5–10 Ma. In the southern zone of the rock basement, the ages of the rock masses are large. The age of the eastern margin of the Andes zone is the Late Jurassic (149–157 Ma), the age of the penetrating migration is the Early Cretaceous (137–145 Ma), and the age of the western margin of the rock basement to the west is 111–136 Ma. The ages of the late magma intrusion are the Late Cretaceous (78–99 Ma) and Paleocene (40–64 Ma), but the rock mass distribution was limited in this period and interspersed in the Late Jurassic and Early Cretaceous rock masses. East of the zone, there are occasional intrusions of Neogene (16–22 Ma) rock masses.[12]

The lithology compositions of the Patagonian rock basement are similar in different locations and feature calc-alkaline features due to orogenic movements around the Pacific.

In addition to this type of lithology, the same orogenic granites appear in the eastern Cordillera zone of Bolivia, where there are granites formed

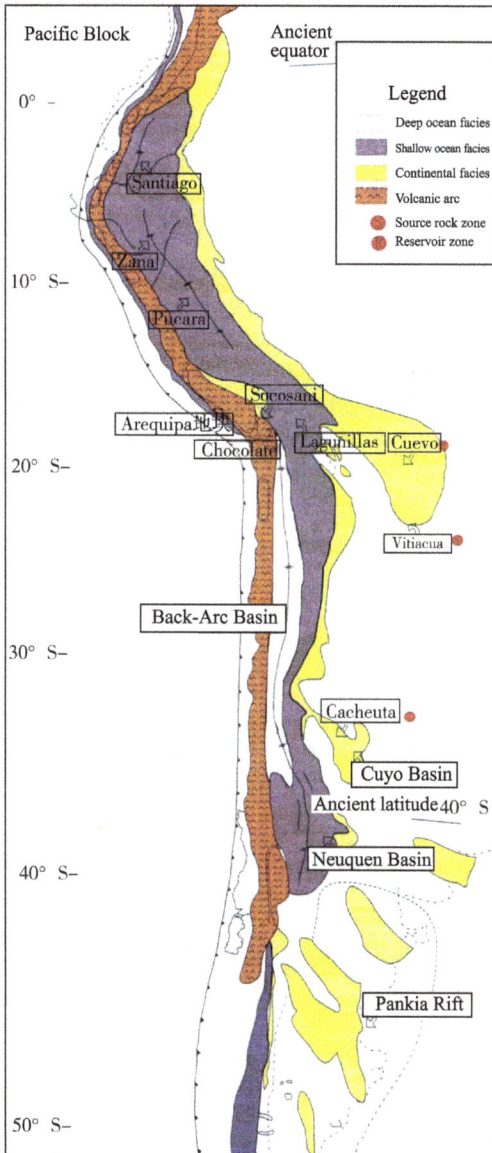

Fig. 8.7.   Paleogeographic diagram of the Triassic–Early Jurassic system on the western coast of South America.

in different periods. The Permian or later granites intruded into the earlier deposit rocks, dominated by acidic granites.

8.4.4 *Volcanic Rocks*

(1) Mesozoic Volcanic Rocks
The Mesozoic volcanic rocks are mainly neutral and calc-alkaline in western South America, dominated by andesites and andesitic-fused tuffs, which are widely distributed in the Andes zone in Peru and Chile. The volcanic activities clearly moved to the east in Chile, and some regions are mainly neutral-acidic. For example, the volcanic rocks at 22°E–29°E are mainly acidic, with high Rb/Sr and K/Rb features, and they are regarded as the pioneers of the Cenozoic "Rhyolite Group", usually known as the "Pioneer of Rhyolite Group" (Fig. 8.8).

(2) Cenozoic Volcanic Rocks
Cenozoic volcanic rocks are well developed in the Andes region. There are more than 600 craters with heights of 5,000–6,800 m at 3,500–4,000 m above the plateau plane in the East Cordillera, and the highest and largest volcanoes in the world are located in this region. The Llullaillalo (6,710 m) and Ojos del Salado (6,880 m) volcanoes on the border between Chile and Argentina are the highest volcanoes in the world.[13]

The Cenozoic volcanic rocks are composed of the early rhyolite group and the late andesite group (Fig. 8.8). The Chile and Bovian rhyolite groups are composed mainly of rhyolite fused tuffs and rhyolites, without folds and with a geological age of 7.1–24 Ma. The lithology of the andesite group is dominated by quartz coarse andesites, quartz andesites, dacites and quartz andesites. The lack of volcanic ash is an important feature of the Cenozoic volcanic rocks in the Andes region.

8.5 S-Shaped Tectonic System in Western Africa

The S-shaped tectonic system spreads over a vast region of western Africa and is composed of the Gabon Basin, Kwanza Basin and Orange Basin.

8.5.1 *Bouguer Gravity Anomaly Features*

The Bouguer gravity anomaly is obtained after the topographic correction, Bouguer correction (height correction and middle layer correction) and

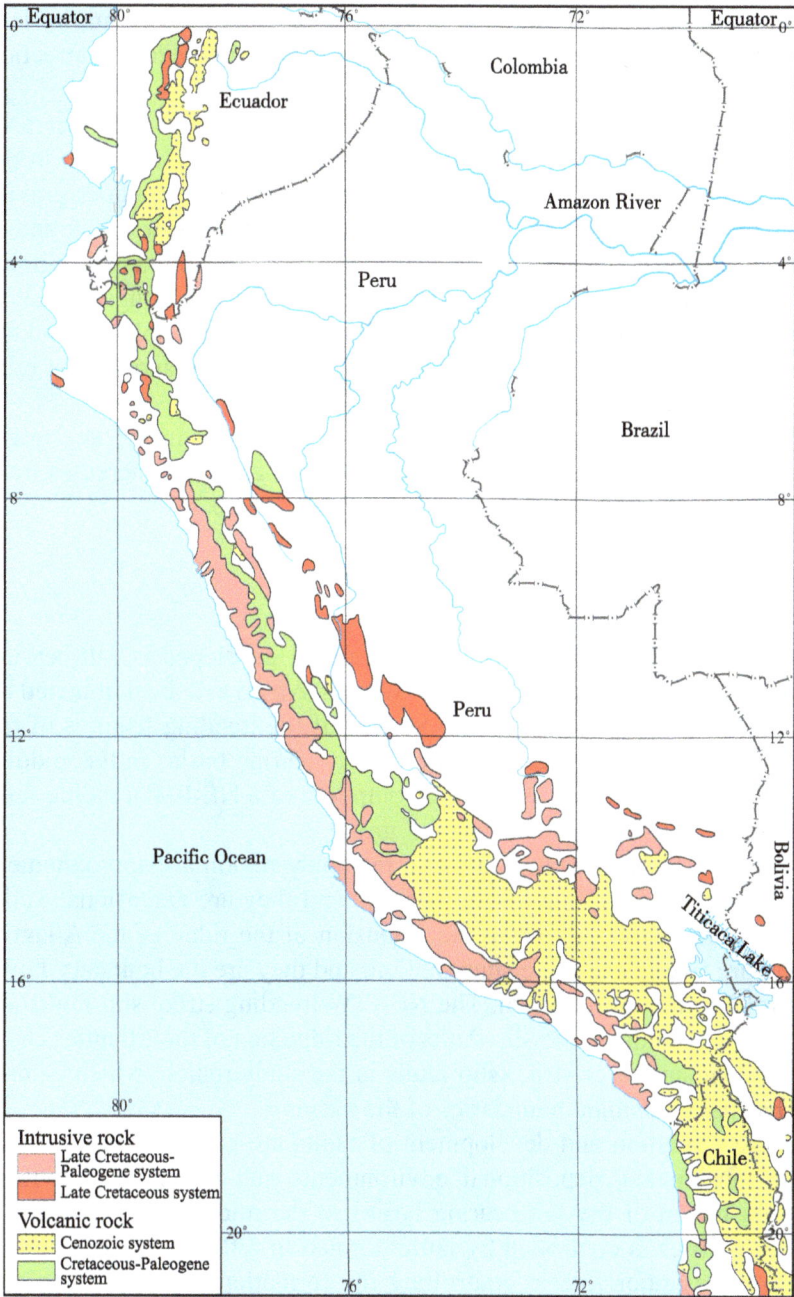

Fig. 8.8. Distribution of Mesozoic and Cenozoic intrusive rocks and volcanic rocks in northwestern South America.

normal field correction (latitude correction) of gravity observations. It is also obtained after the topographic correction and middle layer correction of the free space gravity anomaly (Fig. 8.9).

In the study region, the Bouguer gravity anomaly gradually increases from land to sea. The anomaly is low in the land region and high in the sea region, while the basin is a transition region. The Bouguer gravity anomaly is N-W-trending, nearly N-S-trending and N-E-trending and is caused mainly by changes in the crust thickness. In continental regions, the mantle is sunken, and the crust is thicker, so the Bouguer gravity anomaly is relatively low. In the sea region, the mantle is uplifted, and the oceanic crust is relatively thin, so the Bouguer gravity anomaly is relatively high. Moreover, the basin is a transition region.[14]

The distribution of the Bouguer gravity anomaly in the key target regions shows that the Bouguer gravity anomaly gradually increases from land to sea, which is a reflection of the Moho surface.

8.5.2 *Distribution of Faults*

Many studies have shown that faults are well developed in salt-bearing basins in the middle-southern part of western Africa and are dominated by extensional and strike-slip faults. Studies of the trending features of the faults have shown that the faults in the salt-bearing basins in the middle-southern part of western Africa are composed of a NE-EW-trending fault group and a NNW-NS-trending fault group.[15]

The NNW-NS-trending faults in the study region are approximately parallel to the ridge line of the Atlantic, and they are extensional faults that may have formed during the extension of the ridge of the Atlantic. There are five faults in the study region, and they are the boundary faults controlling the uplift pattern. The NE-E-W-trending strike-slip faults are consistent with the strike-slip faults of the ridge lien of the Atlantic. There are 11 long and large strike-slip faults in the study region, which control the basins or the main boundaries of the basins.

The formation and development of faults are closely related to tectonic movements, depositional environments and magmatic activities. The formation of the salt-bearing basins in the middle-southern part of western Africa is controlled by faults formed in different periods, which also play a major role in controlling the formation of depressions and uplifts in the basin.

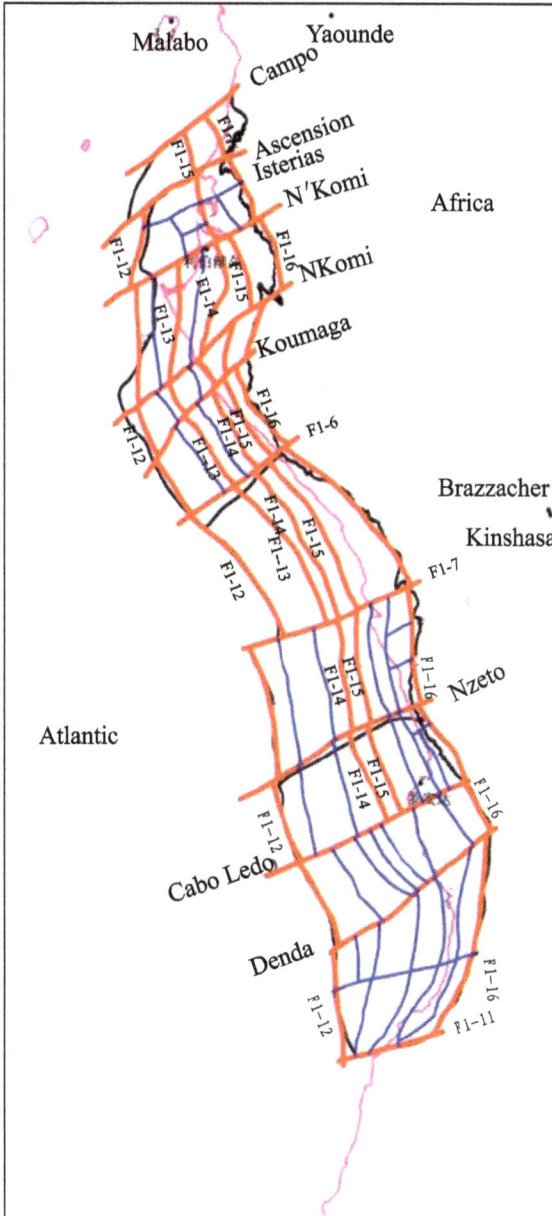

Fig. 8.9.   Bouguer gravity anomaly map. (Unit: mGal)

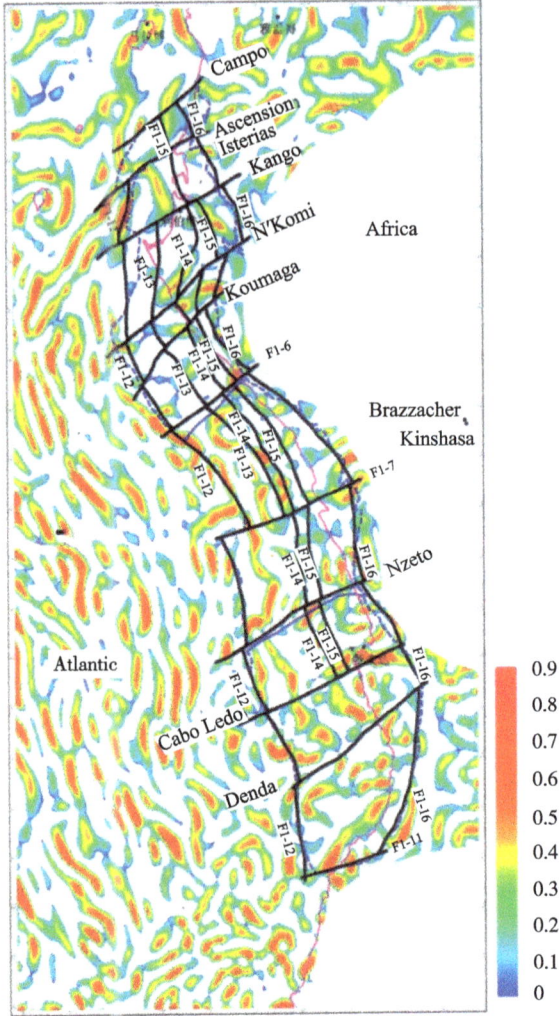

Fig. 8.10.   Polarized magnetic anomaly map. (Unit: nT)

In the study region, there are 16 NNW-NS-trending and NE-EW-trending primary faults (Fig. 8.10) and 30 N-N-W-trending and N-E-E-trending-east secondary faults, with "east stretching and west tensing and N-E-trending strike-slip" features.[16]

Fig. 8.11. Distribution of basins controlled by the 11 S-shaped tectonic system in western Africa.

8.5.3 Distribution of S-Shaped Basins

The Mesozoic and Cenozoic basins are the main components of the S-shaped tectonic system. From north to south, the S-shaped basins on the western margin of Africa include the Lorde Basin (N-E-trending), Senegal Basin (N-S-trending), Sinakri Basin (N-N-W-trending), Abidjan Basin (E-W-trending), Niger Delta Basin (N-W-W-trending), lower Congo Basin (N-N-W-trending), Kwanza Basin (N-S-trending) and Orange Basin (N-W-trending), which together form an S-shaped basin distribution zone (Fig. 8.11).

Overall, this tectonic system has obvious features, and several different-scale fault systems with gravity anomalies and multiple Mesozoic and Cenozoic basins have created an S-shaped tectonic system with obvious tectonic features along the western margin of Africa.[17]

References

1. Shandong Bureau of Geology and Mineral Resources. *Regional Geological Records of Shandong Province*. Beijing: Geological Publishing House, 1991.
2. Jiangsu Bureau of Geology and Mineral Resources. *Regional Geological Records of Jiangsu Province and Shanghai Municipality*. Beijing: Geological Publishing House, 1984.
3. Wang Guolian *et al*. Dragonfly Belt of North China Platform during the Late Carboniferous and Seawater Advance and Retreat Rules. *Editorial Board of Journal of Institute of Geology. Journal of Institute of Geology (#13)*. Beijing: Geological Publishing House, 1989.
4. Wang Hongzhen. Main Stages of Crustal Development in China. *Earth Science*, 1982, (3): 163–186.
5. Wang Xiaofeng. Microstructure Analysis and Stress Estimation of Shaoxing-Jiangshan Fault Zone. *Editorial Board of Journal of Institute of Geology. Journal of Institute of Geology (#12)*. Beijing: Geological Publishing House, 1989.
6. Wang Weixiang. Mechanical Study of Typical Fracture System. *Institute of Geomechanics, Chinese Academy of Geological Sciences. Collection of Geological Mechanics (Part X)*. Beijing: Geological Publishing House, 1995.
7. Wang Zhishun *et al*. Relationship between Cathaysian Structural System and Tancheng–Lujiang Fault Zone in Liaoning, Jiangsu and Anhui Provinces. *Editorial Board of Journal of Institute of Geology. Journal of Institute of Geology (#4)*. Beijing: Geological Publishing House, 1983.

8. Wang Zhishun *et al.* On the Meridional Structural System in the east of Anhui Province and its Significance. *Editorial Board of Journal of Institute of Geology. Journal of Institute of Geology (#5).* Beijing: Geological Publishing House, 1985.
9. Wang Zhishun, Wang Xiaofeng. Mechanical Properties of Tancheng–Lujiang Fault Zone. *Institute of Geomechanics, Chinese Academy of Geological Sciences. Collection of Geological Mechanics (Part XIII).* Beijing: Geological Publishing House, 1988.
10. Wang Zhishun. Huaiyang Epsilon-type System and its Effects on Rock and Ore Control. *Institute of Geomechanics, Chinese Academy of Geological Sciences. Collection of Geological Mechanics (Part IX).* Beijing: Geological Publishing House, 1989.
11. Wang Zhishun, Zhu Dagang. Preliminary Discussion on Metamorphism of Tectonic Dynamics. *Journal of Geomechanics*, 1995, 1(1): 109–111.
12. Wu Anguo *et al.* Basic Characteristics and Rock Control Effects of Structural System in Jiangxi Province. *Institute of Geomechanics, Ministry of Geology and Mineral Resources. Collection of Studies on Provincial Tectonic Systems in China (Volume 1).* Beijing: Geological Publishing House, 1985.
13. Wu Jiang, Chen Zhonghui. Existence of Ancient Bengbu Uplift in Huainan–Huaibei Area. *Coalfield Geology and Exploration*, 1991, (3): 28–34.
14. Wu Leibo, Shen Shumin. Distribution Law of Meridional Tectonic System and Its Geomechanical Significance. *Institute of Geomechanics, Chinese Academy of Geological Sciences. Collection of Geological Mechanics (Part VI).* Beijing: Geological Publishing House, 1982.
15. Tibet Autonomous Region Bureau of Geology and Mineral Resources. *Regional Geological Records of Tibet Autonomous Region.* Beijing: Geological Publishing House, 1993.
16. Xiao Xuchang *et al.* Re-discussion on Plate Tectonics of Qinghai–Tibet Plateau. *Editorial board of periodicals of Chinese Academy of Geological Sciences. Periodicals of Chinese Academy of Geological Sciences (#14).* Beijing: Geological Publishing House, 1986.
17. Gan Kewen, Hu Jianyi. *Atlas of Oil/Gas Containing Basins all over the World.* Beijing: Petroleum Industry Press, 1992.

https://doi.org/10.1142/9789811285561_0009

Chapter 9

Rotation-Torsional Tectonic Systems

Yuzhu Kang, Shuwen Xing, Zongxiu Wang, Zhihong Kang, Yinsheng Ma, Zhijiang Kang, and Zhihu Ling

Abstract

In this chapter, rotation-torsional tectonic systems are introduced, including the rotation-torsional systems in China, the northern Sakhalin geese-shaped systems in northeastern Russia, the broom-shaped systems in the western Indian Ocean and southwestern Pacific, the double-ring compound rotation-torsional systems in Antarctica and the concentric radial systems in the Arctic.

Keywords: Tectonic system; Rotation-torsional; Type

The rotation-torsional tectonic system is formed by rotation-torsional motions centered on a rotation-torsional axis. The rotation-torsional shaft includes upright, inclined and horizontal types. Many rotation-torsional tectonic systems formed by rotation-torsional motions centered on inclined and horizontal shafts occur in the strata after fold movements and are small in scale; they can be observed only on the cross section but cannot be identified on the surface. Rotation-torsional tectonic systems with upright shafts are more common and include three main types: (a) poorly developed broom-shaped tectonic systems; (b) well-developed spiral tectonic systems; and (c) extremely well-developed radiating concentric circular tectonic systems. All tectonic types have the following features:

(1) the central section is composed of cylindrical or semicircular rock blocks or land blocks; (2) the arc-shaped rotation-torsional fault is developed, and the central rocks are divided into a series of arc-shaped rock blocks; (3) these arc-shaped rotation-torsional tension or compression fault surfaces are spread in concentric circles centered on one point; and (4) the extremely well-developed rotation-torsional tectonic systems are accompanied by a series of radial torsional fault surfaces, tectonic initiation zones, rock blocks and land blocks, including broom-shaped, geese-shaped, rotation-torsional, radial and other tectonic systems.

Broom-shaped rotation-torsional tectonic systems are generally composed of several semicircular tectonic traces around the mainstay or vortex (rotation-torsional core) with one end converging and the other end diverging, and the shape resembles a broom. It is a poorly developed but widely distributed rotation-torsional tectonic type, the product of the initial rotation-torsional stage. It can often be seen on both planes and sections and is widely developed in the crust.

The cyclic surface of the broom-shaped tectonics can be a compression-torsional fold or fault zone or a tension-torsional fault zone or dyke group. The former is called compression-torsional broom-shaped tectonics, while the latter is called rotation-torsional broom-shaped tectonics. The rotation-torsional core can be a raised rock block or land block, rock mass, sinking basin, depression, volcanic crater or volcanic neck. The former is called the mainstay, and the latter is called the vortex. The mechanical properties of broom-shaped tectonics are based mainly on the mechanical properties and the relative rotation-torsional direction of the cyclic surface. However, they have their own regularity. For broom-shaped tectonics composed of tensional faults, tension-torsional faults, and dike groups, the internal rotation is always moving in the diverging direction, and the external rotation is always moving in the converging direction. However, for broom-shaped tectonics composed of folds, compression zones or compression-torsional faults, the internal rotation is always moving in the converging direction, and external rotation is always moving in the diverging direction.[1] In short, if the arc-shaped tectonic surfaces of the broom-shaped tectonic system are tensional and rotation-torsional, the rocks around the central section will be twisted from diverging to converging; if those arc-shaped tectonic surfaces are compression-torsional, they will be twisted from converging to diverging (Fig. 9.1).

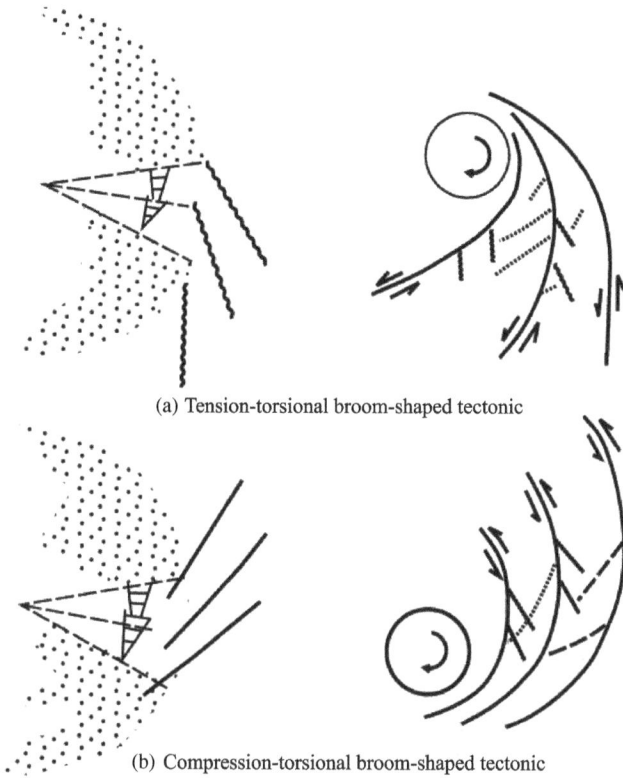

(a) Tension-torsional broom-shaped tectonic

(b) Compression-torsional broom-shaped tectonic

Fig. 9.1.   Schematic diagram of mechanical properties, torsional direction and genetic causes of broom-shaped tectonics.

There are many broom-shaped tectonic examples. Generally, the rotation-torsional cores of large and medium-sized broom-shaped tectonics are upright or nearly upright, reflecting the planar rotation-torsional effect in the region, while some broom-shaped tectonics with nearly horizontal rotation axes are mostly small. Due to the limitation of exposure depth, it is difficult to see or detect some large and medium-sized broom-shaped tectonics with horizontal rotation axes, and these systems need to be studied in detail. Some large long-distance thrust tectonic zones may feature such rotation-torsional tectonics. Based on the available data, the broom-shaped tectonic system can be described only briefly.

9.1 Rotation-Torsional Tectonic Systems in China

9.1.1 *West Gansu Broom-Shaped Tectonic System*

This system spreads in the bordering region of Shaanxi, Gansu, Ningxia and Qinghai in northwestern China, namely, the vast western Gansu region west of Helan Mountain. It is a relatively large-scale broom-shaped tectonic system in China. Its main body is a NWW-SE-trending arc-shaped cyclic fold zone. It starts from the Wushaoling, Minhe and Hekou regions on the west, is compounded with the Qilu–Helan epsilon-shaped tectonic system, and is then inserted into the northern margin of the Qinling Mountains E-W-trending zone in the Baoji region on the southeast, forming three compression-torsional fold zones converging to the northwest, protruding to the northeast, and diverging to the southeast. The fold zones half-surround the Minhe–Hekou Basin to the southwest, forming a vortex as the tensional core.[2] From north to south, the main cyclic fold zones include the Gulang–Tongxin cyclic fold zone, Wushaoling–Liupanshan cyclic fold zone and Yongdeng–Huining cyclic fold zone.

(1) Gulang–Tongxin Cyclic Fold Zone
This zone is a broom-shaped external cyclic fold zone in western Gansu. It starts from Gulang in the west, passes through Changling Mountain and Xiangshan Mountain into the vicinity of Tongxin in the east and turns from E-W-trending to S-E-trending. It is compounded with E-W-trending tectonic zones scattered in the Gansu region, with traces exposed sporadically along the southern margin of the Tengger Desert basin. The section revealed in the Gulang–Peijiaying–Baidunzi region is called the North Changling Mountain fault zone, and the eastern section is well exposed in Ningxia. This zone is composed mainly of the Peijiaying–Changling Mountain compound anticline, Xiaoguan Mountain compound anticline, Gantang–Zhongwei Mountain depression, Qingshuihe, Tongxin, Haiyuan, Guyuan fault depressions and other Mesozoic and Cenozoic basins, as well as the compression and rotation-torsional fault zones on its northern and southern sides (the South Changling Mountain fault and the North Changling Mountain fault). The fault zone is an arc-shaped zone protruding to the northeast and is composed of a series of oblique thrust faults at 15°–80° to the northeast. The newest stratum intersected is the Paleogene–Neogene system, and there are tectonic rock zones with varying widths along the fault zone controlling the distribution of Mesozoic and Cenozoic

continental basins. The folds and thrusts show clockwise rotational-tor-sional trends.

(2) Wushaoling–Liupan Mountain Cyclic Fold Zone

This zone is the main cyclic fold zone of the western Gansu system, with the most obvious and prominent presentation. This zone is a gravity gradient zone, indicating a relatively large effect depth. The western section is inserted into the western wing of the Qilu–Helan front arc west of Wushaoling, passes through Maomao Mountain–Laohu Mountain–Quwu Mountain and Ningxia on the east, connects with the Xihua Mountain, Nanhuashan and Liupan Mountain folds, and is associated with the fault zones. Both sides are limited by fault zones. The middle section is composed of the Maomao Mountain–Laohu Mountain compound anticline, the Xihua Mountain compound anticline and the Liupan Mountain compound anticline. The Cretaceous folds feature right geese-shaped and clockwise rotation-torsion.[3] The Nanhua Mountain and Xihua Mountain zones are tectonic lenses. The continuous broom-shaped tectonics of the arc-shaped fold zone show some morphological features converging to both ends. In the Late Cretaceous, the two rotation-torsional faults turned into compression and rotation-torsional faults and were under compression and uplift. Therefore, Wang's Formation was generally missing in the depression zone and was deposited by the Paleogene continental clastic rock system (fluvial-lacustrine facies, lacustrine facies or fluvial facies).

(3) Yongdeng–Huining Cyclic Fold Zone

This zone starts in the Yongdeng–Tianzhu region along the border between Gansu and Qinghai, which is well exposed in the Huazang Temple–Baiyinchang region on the east, and the section southeast of Baiyin is covered by the Quaternary system. From a geomorphological point of view, its northern boundary may extend from the northeastern margin of the Gaolan block to Gangouyi and to the southwestern side of the Zhuanglang–Zhangjiachuan region, and its southern boundary extends approximately into the Yongdeng–Gaolan Basin and Southwest Huining–Chuankou town regions, as well as the western margin of the Tianshui Basin and Huajialing region further southeast. Overall, the tectonic zone shows poor continuity, and the tectonic lines are intermittent and oblique, converging to the northwest and diverging to the southeast, showing a NNW-SE-trending arc protruding to the north. Most sections of the fault zone are oblique to the north, with obvious geomorphological features. They have good continuity in the Ganluchi–Baiyin–Hining region and are composed of several faults. They are the tectonic zones that separate the Meso-New Proterozoic and Jingyuan–Jingning Mesozoic and Cenozoic troughs. The latest strata intersected are the Tertiary system. Along the fault zone, fault gouges, tectonic lenses, scratches and scratch surfaces are commonly shown in clockwise torsion. There are many rotation-torsional signs in Xigongyi, Huining and its southeastern margin.[4]

9.1.2 Hebei–Shandong Broom-Shaped Tectonic System

There is a large broom-shaped tectonic system under the Quaternary system on the North China Plain. Its main cyclic fold zones are the Renqiu Depression, Cangzhou Uplift, Huanghua Depression, Chengning Uplift, Jiyang Depression, etc., diverging to the northeast and converging to the southwest. Its mainstay is the West Shandong Uplift. This system spreads in the region north of the Qinling Mountains E-W-trending zone. It is the product of the relative rotation-torsional movement between the western Shandong rock block and the Hebei–Shandong Plain in the Himalayas, forming an arc composed of uplifts and depressions with the primary cyclic layers protruding to the northwest and semicircular distribution around the West Shandong mainstay. The primary uplifts and depressions are again controlled by secondary depressions, uplifts and their arc-shaped faults. All cyclic layers have obviously regular geological features with large N-S-trending differences and few E-W-trending differences. The lower-level rotation-torsional tectonics in the cyclic layers are very developed, and the main faults controlling the depressions are generally compression-torsional and closed. Therefore, the Central Hebei, Huanghua and Jiyang arc-shaped depression zones (three broom-shaped negative cyclic tectonic zones) control the formation and distribution of three large-scale compound hydrocarbon accumulation zones in the central and northern sections of the North China Plain (Fig. 9.2).

The Hebei–Shandong broom-shaped tectonic system has undergone two stages of development. At the end of the Mesozoic, a tension-torsional broom-shaped tectonic system converging to the external torsion direction and diverging to the internal torsion direction formed. Therefore, the Paleogene system was deposited in the depression cycle zone (the Cretaceous system was also deposited in some sections) and suffered

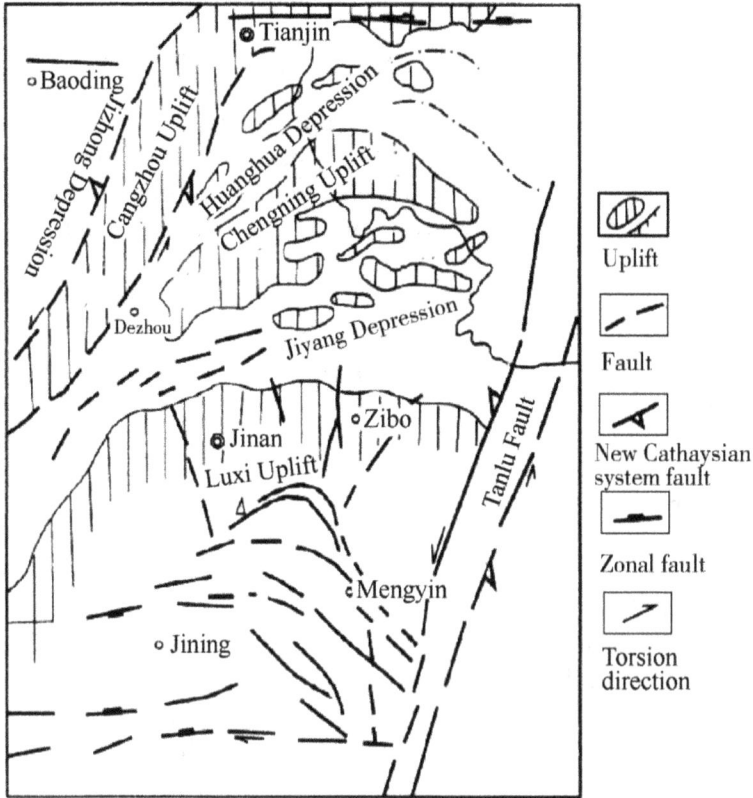

Fig. 9.2. Schematic diagram of the Hebei-Shandong broom-shaped tectonic system (simplified).

the effect of tectonic deformation, and the Neogene system was uncon-formable on it. Since the Neogene, due to changes in the regional stress field, a compression-torsional broom-shaped tectonic system diverging to the outer torsion direction and converging to the inner torsion direction was formed, namely, the current Hebei–Shandong broom-shaped tectonic system. Since the Neogene, the North China Plain has depressed greatly, and the thicknesses of the Neogene and Quaternary deposits have increased. For example, the thicknesses are 5,500–6,500 m in the Dongying Depression, approximately 2,500 m in the Paleogene system, and more than 3,500 m in the Cenozoic system in other depressions.[5]

An analysis of nine petroliferous basins in the North China depression region from petroleum departments indicates that secondary rotation-torsional tectonic systems with different scales, different levels, and different stress modes have widely developed in all cycle zones of the Hebei–Shandong broom-shaped tectonic system, including broom-shaped, herringbone-shaped, ξ-shaped, checkerboard-shaped, ring-shaped, radial and other rotation-torsional tectonic types. They exhibit a certain regularity in their spatial distribution. The broom-shaped, ring-shaped and other small torsion tectonic types are developed mainly in the inner cycle layer adjacent to the torsional core, while the ξ-shaped, herringbone-shaped and other torsion tectonic types are developed mainly in the middle and outer cyclic layers away from the torsional core. Their scales and levels are very different; the largest secondary rotation-torsional system has an area of up to 220,000 km^2, and the small ones can be seen in the draft.

The Hebei–Shandong broom-shaped tectonic system is based on the tension-torsional broom-shaped tectonic system formed in the early Mesozoic and transformed under the stress modes that changed in the late Cenozoic. It has developed synchronously with the West Shandong broom-shaped tectonic system. Because of their different convergence directions, the mechanical properties in the early and late phases are exactly opposite. In the early period, it had a genetic relation with the N-N-E-trending system. In the late phase, it might have been related to the stress field of the N-S-trending tectonic system that suffered N-S-trending compression in this region; that is, it was the product of the N-S-trending zone under uneven compression and rotation-torsion in the region. The former has obvious control over the formation and deposition of depression troughs, while the latter has important control over the regional tectonic deformation and the migration and accumulation of oil and gas sources. These features are noted in studies of the migration, accumulation and distribution of oil and gas sources in the N-N-E-trending troughs in eastern China.

The latitudes of the West Shandong and Hebei–Shandong broom-shaped tectonic systems are equivalent to those of the West Gansu broom-shaped tectonic system. They are spread between the Yin Mountain–Tian Mountain E-W-trending zone and the Qinling Mountains–Kunlun E-W-trending zone. They correspond to each other, without similar combination forms, early mechanical properties, formation and development phases and movement directions. They formed during strong activities from the Late Cretaceous to the Paleogene,[6] reflecting the southward movement of the central section of Mainland China and the northward rotation-torsional movement of the eastern and western sections of Mainland China, resulting in typical and partial rotation-torsion along the paleo-continental masses.

9.1.3 *Yakela–Luntai Broom-Shaped Tectonic Zone in the Tarim Basin*

The zone spreads north of the Shaya uplift to the north of the Tarim Basin, starting from Xinhe County to the west and extending east of Luntai to the east, with a length of approximately 200 km and a width of 20–80 km, similar to a broom converging from west to east. The tectonic zone is controlled mainly by strike-slip faults, namely, the Yaha fault in the north, the South Yahan fault in the middle and the Luntai fault in the south. Multiple geese-shaped faults and fold zones have formed on the upper and lower walls of the faults (Fig. 9.3). The broom-shaped tectonic system formed mainly in the Mesozoic and Cenozoic, and more than 20 oil and gas fields have been discovered in the upper and lower walls of the main faults.

9.1.4 *Central Tarim Broom-Shaped Tectonic Zone*

The Central Tarim broom-shaped tectonic zone spreads in the Khattak uplift in the middle of the E-W-trending uplift zone of the Tarim Basin.

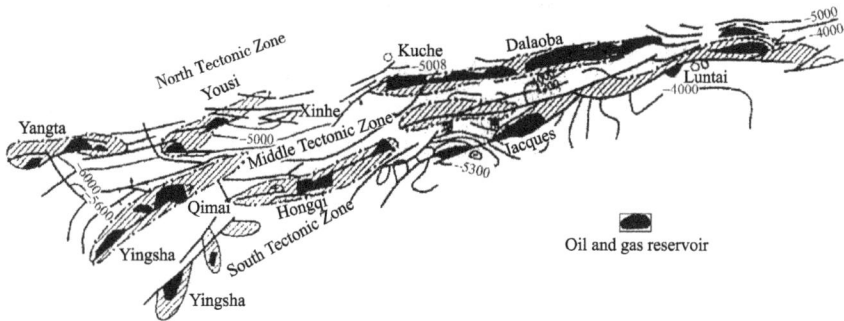

Fig. 9.3. Distribution of oil and gas fields controlled by the Yakra broom-shaped tectonic zone.

It is a broom-shaped tectonic system diverging to the northwest and converging to the southeast.

Many fault tectonic zones are developed in the Central Tarim region. Under the control of various factors, the fault activities obviously differ between different fault zones or in different sections of the same fault zone. Based on seismic tectonic analysis, the forms and sections of these tectonic zones were first studied, and their evolutionary causes were inferred. By studying the oil and gas accumulation conditions of the main tectonic zones and analyzing the role of faults in the accumulation process, the relationship between the broom-shaped tectonic system and oil and gas accumulation was explored.

The results of the regional profile study show that under regional compression stress in the Late Ordovician, the tensional environment transformed into a compression environment in the Central Tarim region, resulting in the reverse of tectonic features, the Central Tarim No. 1 and No. 3 fault zones. The Tangbei fault system and the Tumxuke fault systems have all been formed; the No. 2 fault zone is a large-scale compression-derived fault zone with the properties of a transition zone. The E-W-trending and E-W-trending ancient faults of the Central Tarim uplift initially formed at that time.

After the early Caledonian large-scale tensional movement, the crust in this region changed from stretching to shrinking at the end of the Middle Ordovician, the tectonic pattern of the Tarim platform was E-W-trending, and the Central Tarim E-W-trending uplift zone played a framing role in the tectonic pattern of the platform region. Due to the inhomogeneity of the N-S-trending compressional stress, the western section of the Central Tarim uplift is wider, and the axis of the low uplift presents finger-like diverging to the west. The early rotation-torsional tectonic zone associated with the Central Tarim No. 1 fault to the north, the thrust imbricated fault zone diverging west of the central Tarim No. 3 fault to the south, and the associated thrust anticlines generally formed in the same period. The Silurian and Devonian systems are basically overlying deposits on a paleoeroded background, with the thickest deposits in the southern and northern depressions of the Central Tarim low uplift.[7] At the end of the Devonian, the early tectonic pattern was further strengthened by the overall uplift on the East Tarim region, the paleo-uplift was further enlarged, and the Silurian, Devonian and Upper Ordovician systems were again eroded. The tectonic deformation and transformation were the strongest in the middle east of the central Tarim low uplift, and the early tectonic zones were further complicated and finalized by thrusting, uplifting and erosion by the strengthening of faults. The southern and northern wings of the central Tarim low uplift and the Gucheng nose uplift were relatively less affected. Due to the reactivation of the central Tarim No. 2 fault, the ancient topography was basically flattened in the central Tarim region due to uplift and erosion in the Devonian. During the early Carboniferous deposition, the central Tarim Qianshan Mountain zone formed a southern barrier in the "Island Chain" form. To date, the central Tarim broom-shaped tectonic pattern has basically finalized. After the Carboniferous, the central Tarim region was basically stable, without the overall uplift and depression, and only some faults were active in the central and western parts of the uplift. Due to the influence of the late Hercynian movement, basic magma emplacements occurred along the main fault of the broom-shaped tectonic zone. In the Himalayas, due to the influence of Altyn's left-handed rotation-torsional stress field, the central Tarim broom-shaped tectonics were finalized.

The formation and evolution of fault tectonics are controlled by the regional tectonic pattern in the central Tarim region. The fault tectonics are generally distributed in a broom-shaped plane in the central Tarim region. There is a large-scale compound anticline, and the two wings form a thrust fault system, which is generally a large flower-like fault tectonic on the cross-section. The combination types of the fault tectonic zones include flower-like, Y-shaped back thrust, imbricated thrust, and hedge. The backbone fault is basically of the base-entangled type, and the nature of the fault is mostly compression and rotation-torsional. Most faults were formed in the Caledonian and finalized in the pre-Carboniferous. The active Late Hercynian faults are concentrated west of the uplift and were basically stable after the Permian.[8] These features of the central Tarim broom-shaped tectonic system formed during the regional geological tectonic evolution process (Fig. 9.4).

9.1.5 *Geese-Shaped Tectonic System in the Tarim Basin*

This tectonic system is widely distributed. For instance, the Gongtake geese-shaped tectonic zone north of the Tarim Basin is located west of the Kuqa Depression and the Paleogene–Neogene geese-shaped tectonic zone formed under the control of the N-W-trending system in the Himalayas.

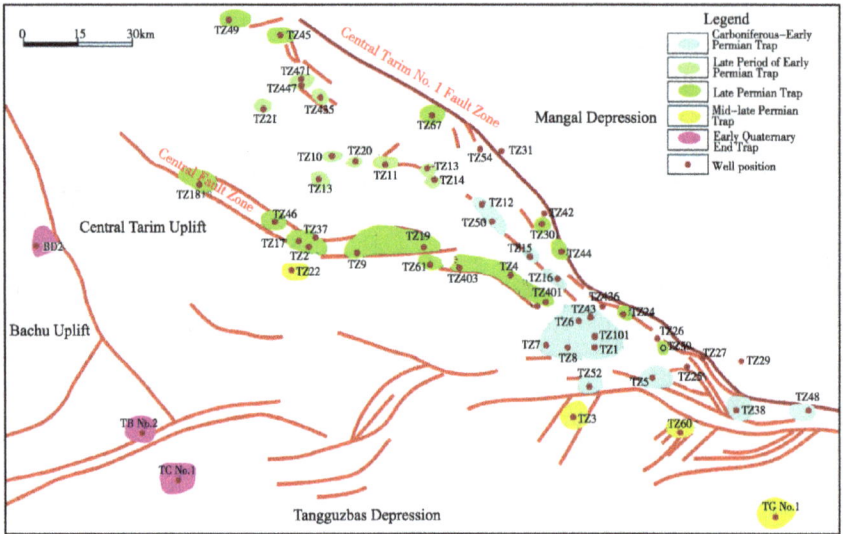

Fig. 9.4.   Layout of the main tectonic traps in the Central Tarim Carboniferous system in the initial formation period.

9.1.6 *Akekule Rotation-Torsional Tectonic System in the Tarim Basin*

This rotation-torsional tectonic system spreads on the Akekule uplift in the middle section of the Shaya uplift on the northern part of the Tarim Basin. It formed mainly in the Hercynian and is composed of two cyclic tectonic zones and vortices, which play an important role in controlling the distribution of oil and gas fields in this region.

The formation process of the Akekule rotation-torsional tectonic system is as follows.

(1) Sinian–middle Ordovician

A set of neritic carbonate rocks were deposited in the Craton Basin, and the tectonic activity was relatively weak in this region. The N-S-trending compressions occurred at the end of the middle Ordovician, resulting in the formation of uplifts, the Akkumu fault zone and the Sangtam fault zone in this region.

(2) Late Ordovician–Silurian

The Tarim Platform was in a compressional environment. During the development period of the flexible basin, the Akekule N-E-trending anticline uplift was formed, and the Akkumu and Sangtam faults were again uplifted, resulting in strong erosion in the region.

(3) Devonian
Due to the continuous regional compression stress, the region was uplifted again, and fault activity increased, resulting in the formation of the Akekumu and Akekule back-thrust fault tectonic zones.

(4) Carboniferous–Permian
Due to the stretching and depression of platforms during the early middle Carboniferous, deposits may have formed in this region, forming a set of neritic carbonate rocks. They were uplifted again in the Late Carboniferous, and the upper Carboniferous system was missing. In the early Permian, some volcanic eruptions also occurred in this region, forming a set of pyroclastic rock formations, which were again uplifted and strongly eroded in the late Permian. At that time, the Akekule rotation-torsional tectonic zone was basically finalized, and it played an important role in controlling the distribution of oil and gas fields in this region (Fig. 9.5).

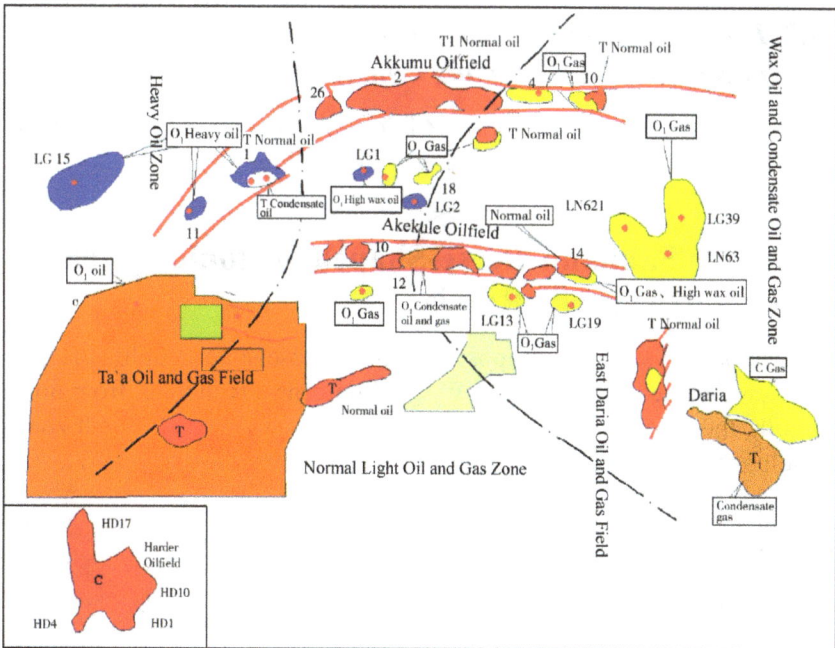

Fig. 9.5.   Distribution of Tahe oil and gas fields and surrounding oil and gas fields in the northern Tarim Basin.

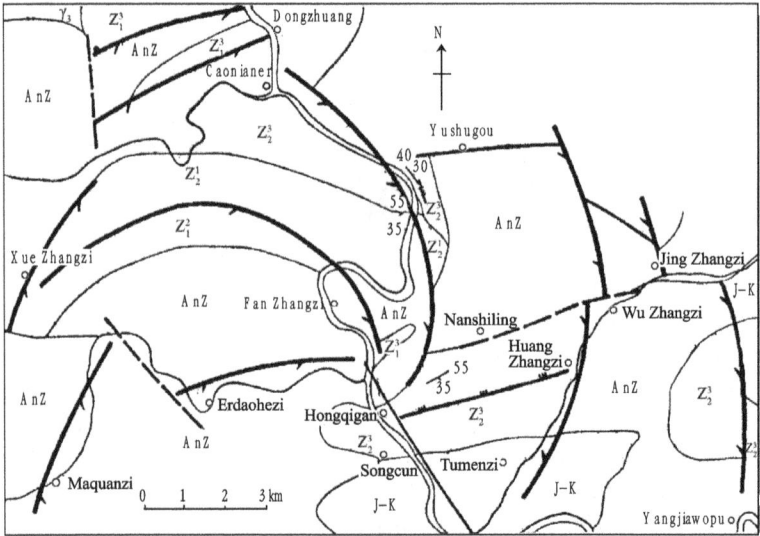

Fig. 9.6. Schematic diagram of the Hongqigan radial tectonic zone in southeastern Qinglong County in Hebei Province.

9.1.7 *Hongqigan Rotation-Torsional Radial Tectonic System in Southeastern Qinglong County in Hebei Province*

The tectonic system is composed of four radial fault zones (Erdaohe, Fanzhangzi, Caonianer and Wuzhangzi faults) (Fig. 9.6).

9.2 North Sakhalin Geese-Shaped Tectonic System in Northeastern Russia

This tectonic system is composed of four geese-shaped uplift tectonic zones sandwiched by three N-N-W syncline zones, including, from north to south, the Odopkin Sea, East Odopkin, Odopkin and Nayi uplift zones as well as the Tapnin, Chayvo and Lensky syncline zones (Fig. 9.7).

9.3 West Indo-Ocean Broom-Shaped Tectonic System

The tectonic system is composed of three ridges self-surrounding the island of Madagascar. The first is the Seychelles ridge located in the

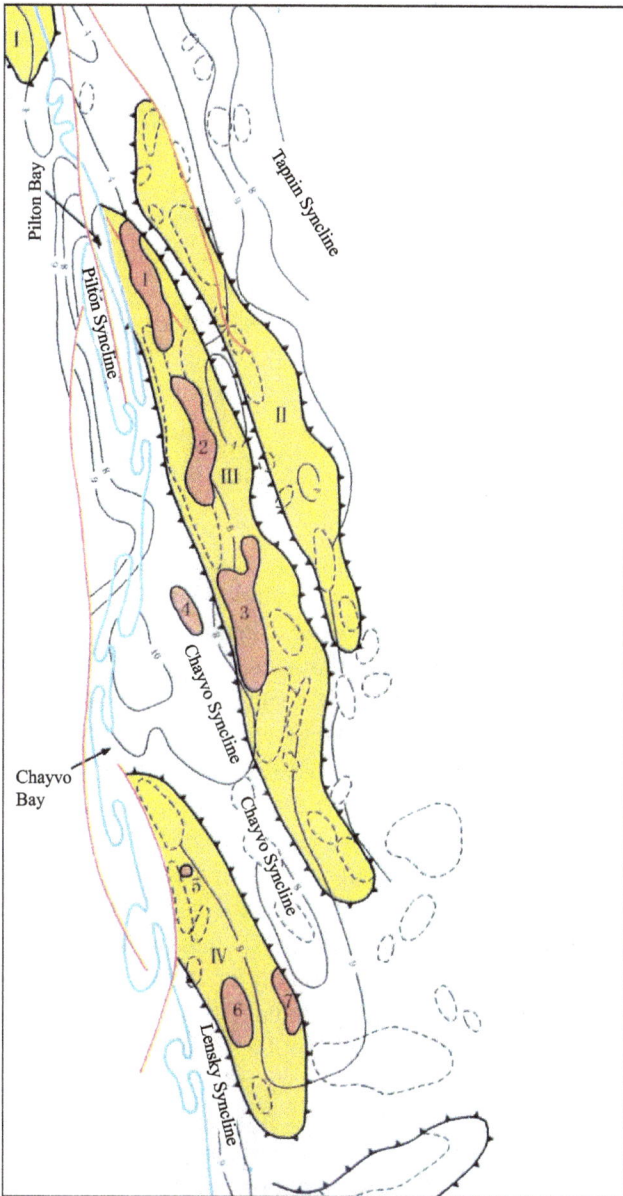

Fig. 9.7.   Distribution of the geese-shaped tectonic system in North Sakhalin, Russia.

Notes: Oil and gas fields: (1) Odopkin Sea; (2) Pirtun; (3) Alcuton-Dakin; (4) Chayvo; (5) Vining; (6) Lensky; (7) Keelin.

Tectonic zones: I — Espenbiere; II — East Odopkin; III — Odopkin; IV — Nayi.

westernmost region (Seychelles Seawall) at 52°E–62°E and 3°N–20°N, and there are 4,000 m deep trenches in a crescent shape on both sides. The second is the Karlsberg ridge (Karsberg Longwall), which is northeast of the first zone. It is relatively narrow, with deep seas on both sides, and there is a crack in the middle, filled with basalts. The third is the Lagosdev ridge; its northern end enters the Indian Peninsula, and its southern end is nearly perpendicular to the Karsberg ridge, with a slight curve. These arc-shaped submarine uplift zones, including the Madagascar zone, form the Southwest Indian Ocean broom-shaped tectonic system, reflecting the horizontal differential rotation movements in the Indian Ocean and the African continent in the Mesozoic and Cenozoic.[9]

9.4 Southwest Pacific Broom-Shaped Tectonic System

This tectonic system is composed of a series of arc-shaped islands in the southwestern Pacific, including a system of arc-shaped islands closest to the Australian mainland, such as the Rennell Islands, New Caledonia and Norfolk Islands, as well as the Auckland Peninsula of New Zealand, i.e., the North Island. This system of islands and peninsulas is comparable to the first Australian arc called Suzie (named after an Austrian scholar). Northeast of this arc is another system of arc islands, which include the Solomon Islands, New Georgia Islands and New Hebrides Islands. Farther north and northeast, there is another system of arc-shaped islands composed of the Caroline Islands, Cusaier Island and Nauru Island. Farther northeast is another system of arc-shaped islands, with the northern section divided into an eastern branch (Meadak Islands) and a western branch (Lalik Islands), which are called the Marshall Islands. This tectonic system extends to the Gilbert Islands, Tuvalu Islands, and Fiji Islands to the south. This system of arc-shaped islands is equivalent to the second Australian arc called by Suzie. The southeastern ends of the above four arc-shaped islands show a tendency to insert into the Tonga–Kemadek–New Zealand uplift zone, and there are some northward movement signs of the western side of the uplift zone corresponding to its eastern side. Some researchers believe that the staggering distance was up to 300 km. These four arc-shaped islands form a broom-shaped tectonic system composed of large rotation-torsional faults. Based on information about the New Zealand Alps fault, the internal rotation layer of this broom-shaped tensional tectonic system is moving counterclockwise, while the external rotation layer, the Alps fault, is staggered clockwise. The staggering distance was 480 km since the Jurassic and 18 km only in the Quaternary. Therefore, this broom-shaped tensional tectonic system originated in the Jurassic and is still active today. It is an earthquake zone.

9.5 Antarctic Double-Ring-Compounded Rotation-Shearing Tectonic System

The Antarctic double-ring-compounded rotation-shearing tectonic system differs from the Arctic rotation-shearing tectonic system as follows: (a) It is a region with the Antarctic continent as the core and surrounded by the Southern Ocean, with an area of 1400×104 km^2. It is known as the 7th continent and is the highest continent in the world. It is located at an altitude of 2,350 m and is nearly entirely covered by glaciers with an average thickness of 1,880 m and a maximum thickness of more than 4,000 m. The point exposed at the southern end of the Earth's rotation axis is the center of rotation motion. (b) It is a region controlled by the radial horizontal force, rotation force and ocean rift pushing and pulling force generated by the superimposed motion of the Earth's self-strike motion and revolution motion. (c) It has double-ring-compounded tectonic features. A rift-like rotation-shearing tectonic zone surrounded by the Southern Ocean is developed on the periphery of the Antarctic continent as indicated above. The rift system forms the first ring surrounding Antarctica. One strike thrust points to the north and pushes the surrounding geological masses in the direction of the North Pole (it must be related to the current continents concentrated in the Northern Hemisphere). The other strike thrust points to the south and pushed the geological masses inside the ring fault system in the direction of the South Pole. Under relative compression from these thrusts and the radial centrifugal force at the center of the pole, the second and third concentric ring compression tectonics formed, including a series of arc-shaped thrusts and folds as well as matching radial fault tectonics exposed in many regions. (d) In Antarctica, due to the interference of other tectonic systems, there are an S-shaped orogenic zone across Antarctica, S-shaped rift systems across western Antarctica, and Meso-Cenozoic S-shaped orogenic zones in Antarctic Peninsula–West Antarctica, showing a more colorful tectonic pattern in Antarctica.[10]

9.6 Arctic Concentric Radial Rotation-Shearing Tectonic System

The geological features of the concentric radial rotation-shearing tectonic system in the Arctic and subarctic regions are completely different from those in the Antarctic in the following ways: (a) The Arctic region is mostly covered by the Arctic Ocean, and the North American continent and the Eurasian continent converge toward the Arctic region in a circle. The Arctic Ocean is the smallest of the four oceans in the world and is elliptical, with an area of 1310×104 km^2 (equivalent to $1/14$ of the Pacific Ocean), an average depth of 1,200 m, and a maximum depth of 5,449 m in the Nansen Basin. The topography of the ocean floor is quite complex, and three ridges [Arctic Ocean (Nansen) Ridge, Lomonosov Ridge and Mendeleev Ridge] and two large basins (Eurasia Basin and Canadian Basin) are arranged along the minor axis. The continental shelf is very wide, with an area of approximately 400×104 km^2, accounting for $1/3$ of the entire Arctic Ocean. (b) The point exposed at the northern end of the Earth's rotation axis, the North Pole, is also located at the center of rotation motion, and the crustal superimposition movement initiated by the Earth's rotation and revolution generates the deviating force and the pushing force to drive the continental blocks to the North Pole, resulting in the formation of annular and radial faults and fold tectonics. (c) The Arctic compression concentric radial rotation tectonic zone is similar to a tectonic system formed by the force field generated by driving a rotating cone into an object. When the rotating cone is driven under the combined force, it is not only rotated but also drilled forward. Therefore, it must be subject to a counter force, thus forming a distinct tectonic trace and pattern. (d) The concentric tectonic traces in this tectonic system have a compression feature in faults or other tectonic forms, and the corresponding radial tectonic traces have a compression-torsional feature in faults or other tectonic forms.

The outermost ring of the Arctic tectonic system is the stable northern Precambrian craton block system, which fully reflects these features as follows: (a) The block system is a concentric circle surrounding the Arctic rotation center; (b) the concentric tectonic traces have a compression feature in the block system; (c) the radial tectonic zones, i.e., Ural fold fault zone, Scandinavian fold fault zone, East Greenland fault zone, Yenisei River fault zone, upper Verkhoc fold zone, have compression and rotation-torsional features; and (d) the blocks between the concentric

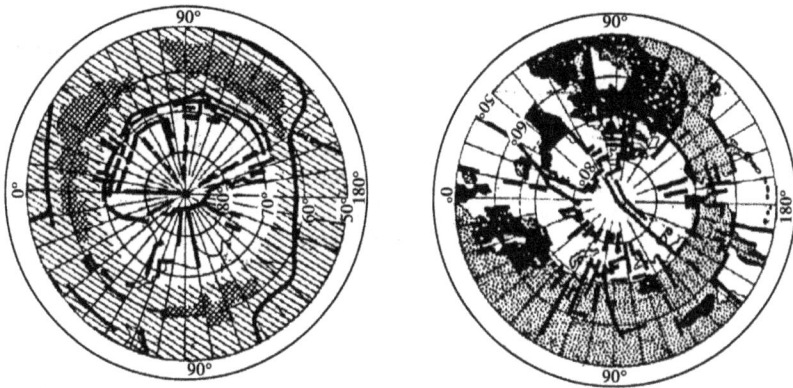

Fig. 9.8.   Schematic diagram of the Arctic concentric radial rotation-shearing tectonic system.

circles and the radial tectonic zones are mostly trapezoidal or square, with the short side of the trapezoidal blocks located on the side of the rotation center and the long side located on the periphery, showing the wedging effect of the trapezoid blocks in the formation of the concentric radial tectonic zones.[11] The Arctic concentric radial compression tectonic system is shown below (Fig. 9.8).

References

1. Gan Kewen, Li Guoyu, Zhang Liangcheng *et al. Atlas of Petroliferous Basins all over the World*. Beijing: Petroleum Industry Press, 1982.
2. Guan Zengmiao. *African Oil and Gas Resources and Exploration*. Beijing: Petroleum Industry Press, 2007.
3. Hu Jianyi, Xu Shubao *et al. Description of Oil and Gas Prospects Evaluation Map in the Northeast Asia*. Beijing: Petroleum Industry Press, 1996.
4. Jin Zhijun, Yin Jinyin. *Petroleum Geology Characteristics and Oil and Gas Distribution in Asia*. Beijing: China Petrochemical Press, 1997.
5. Li Chunyu *et al. Asian Geotectonic Map (1: 8,000,000)*. Beijing: Geological Publishing House, 1984.
6. Li Guoyu, Jin Zhijun. *Atlas of Petroliferous Basins all over the World (Volumes 1 and 2)*. Beijing: Petroleum Industry Press, 2005.
7. Li Guoyu. *Petroleum Geology all over the World*. Beijing: Petroleum Industry Press, 2003.
8. Liu Luofu, Zhu Yixiu. *Geological Characteristics of Oil and Gas in the Pre-Caspian Basin and Central Asia*. Beijing: China Petrochemical Press, 2007.

9. Ren Jishun. *1: 500000 Geotectonic Map and Descriptions of China and its Neighboring Regions: A Global View of China's Geotectonics.* Beijing: Geological Publishing House, 1999.
10. Tong Xiaoguang, Dou Lirong, Tian Zuoji *et al. Research on China's Transnational Oil and Gas Exploration and Development Strategy in the early 21st Century.* Beijing: Petroleum Industry Press, 2003.
11. Tong Xiaoguang, Guan Zengmiao. *Atlas of World Petroleum Exploration and Development.* Beijing: Petroleum Industry Press, 2004.

Chapter 10

Evolutionary Features and Compound Relationships of Tectonic Systems

Yuzhu Kang, Shuwen Xing, Zongxiu Wang, Zhihong Kang, Yinsheng Ma, Zhijiang Kang, and Zhihu Ling

Abstract

In this chapter, the evolutionary features and compound relationships of tectonic systems are introduced. Five evolutionary features are proposed as follows: phase, inheritance, migration, difference and conversion, showing the complexity of all tectonic systems. The genetic evolution of various blocks is controlled by tectonic systems, while the formation and evolution of tectonic systems are controlled and affected by various blocks. Their interactions have created the current global tectonic pattern and land–sea changes and evolution.

Keywords: Tectonic system; Evolution; Compound

10.1 Evolutionary Features of the Tectonic System

The main features of the tectonic system in its formation, development and evolution process can be preliminarily summarized into five aspects as follows: phase, inheritance, migration, difference and conversion.

(1) Phase

The formation of each tectonic system was not completed in one period and one movement. The formation and development of the tectonic system were driven by the multirotation feature of regional tectonic movements. For example, the Shaya–Kuruktag paleo-uplift of the E-W-trending system was an E-W-trending depression zone in the Sinian-middle Ordovician, uplifted in the Late-middle Ordovician, rapidly uplifted and basically finalized in the Silurian–Devonian, and was still under uplift in the Carboniferous–Permian. Some regions were eroded, and the Mesozoic and Cenozoic Shayazhan uplift was mostly depressed and covered.[1]

In addition, some major tectonic masses, such as the Junggar Basin, Ordos Basin and Sichuan Basin, all show phasal evolutionary features.

Another example is the N-W-trending system. The Bachu uplift was initiated at the end of the middle Ordovician and was further uplifted in the Late Ordovician–Devonian, resulting in the erosion of the upper Ordovician–Devonian system. The central Tarim Basin uplift was not active in the Carboniferous and was covered by the Carboniferous system, but the Bachu uplift was uplifted continuously until the Himalayan.

(2) Inheritance

The Tarim and its surrounding tectonic systems have experienced at least two phases of inhibitive development and evolution, while the E-W-trending and northwestern systems have experienced at least four inhibitive phases of development and evolution, i.e., the end of the middle Ordovician, the end of the Silurian, the end of the Devonian and the end of the Carboniferous–Permian. The abovementioned tectonic systems show inheritance features under the actions of multiphasal subregional tectonic movements.

(3) Migration

The evolution of each tectonic system is not uniform, and tectonic activity has migration features in different development periods. The deposit center of the N-W-trending system was located in the Manjiaer region east of the Tarim Basin in the Early Paleozoic but migrated to the Yecheng region in the west in the Late Paleozoic. Another example is the Kuche depression in the Tian Mountain E-W-trending system. The main depression center was located in the basin in the Triassic–Jurassic but migrated to the Kashgar depression of the E-W-trending Kunlun system in the Cenozoic.

The migration features of deposit centers and depression centers are shown fully in the surface deposit features in northwestern China in various periods. For example, as shown on the deposit map, the large-scale deposits were distributed in the northwestern region and only missing in the east Tarim block and the middle and west of the Hexi Corridor region in the Permian, but were greatly reduced in the early Mesozoic–Triassic and only distributed in the Junggar, Santang Lake, Tuha, northeastern Tarim and southeastern Hexi corridor regions (Fig. 10.1).

Era	Geological Age	Tectonic Stage	Block Movement	Tectonic Stage and Block Movement	Tectonic System Evolution	Basin Formation
Cenozoic	Quaternary					
Cenozoic	Neogene	Himalayan Indosinian Tectonic Stage	Inland Basin Develop-ment Stage	Late Himalayan Movement		Foreland Basin
Cenozoic	Paleogene	Himalayan Indosinian Tectonic Stage	Inland Basin Develop-ment Stage	Middle Himalayan Movement / Early Himalayan Movement		Foreland Basin
Mesozoic	Cretaceous	Himalayan Indosinian Tectonic Stage	Inland Basin Develop-ment Stage	Late Yanshan Movement		Foreland Basin
Mesozoic	Jurassic			Middle Yanshan Movement / Early Yanshan Movement		Foreland Basin
Mesozoic	Triassic			Indochina Movement		Foreland Basin
Late Paleozoic	Permian	Hercynian Tectonic Stage	Block Opening and Closing Develop-ment Stage	End Hercynian Movement / Late Hercynian Movement		Craton depression basin
Late Paleozoic	Carboniferous	Hercynian Tectonic Stage	Block Opening and Closing Develop-ment Stage	Middle Hercynian Movement		Craton depression basin
Late Paleozoic	Devonian	Hercynian Tectonic Stage	Block Opening and Closing Develop-ment Stage	Early Hercynian Movement		Craton torsion depression basin
Early Paleozoic	Silurian	Caledonian Tectonic Stage	Ancient Continen-tal Block Splitting and Rifting Stage	Late Caledonian Movement		Craton torsion depression basin
Early Paleozoic	Ordovician	Caledonian Tectonic Stage	Ancient Continen-tal Block Splitting and Rifting Stage	Middle Caledonian Movement		
Early Paleozoic	Cambrian	Caledonian Tectonic Stage	Ancient Continen-tal Block Splitting and Rifting Stage	Middle Caledonian Movement		Craton rift basin
Neoproterozoic	Sinian	Caledonian Tectonic Stage	Ancient Continen-tal Block Splitting and Rifting Stage	Early Caledonian (Xingkai) Movement		Craton rift basin
Neoproterozoic	Qingbaikou Period					
Mesoproterozoic	Jixian Period	Jinning Stage	Continen-tal Crust Formation Stage	Jinning (Talimu) Movement	E-W-trending system | N-W-trending system | N-E-trending system | Reverse S-shaped system | N-N-E-trending system	
Mesoproterozoic	Nankou period	Jinning Stage	Continen-tal Crust Formation Stage	Jinning (Talimu) Movement		
Mesoproterozoic	Changcheng Period		Continen-tal Crust Formation Stage			
Paleoproterozoic	Lvtuo Period	Luliang Stage	Continen-tal Crust Formation Stage	Luliang (Xingdi) Movement	Zhungeer block formation	
Paleoproterozoic	Wutai Period	Luliang Stage		Wutai Movement	North Tarim and Alxa block formation / South Tarim block formation	
Archaic	Fuping Period	Fuping Stage	Land Nucleus Formation Stage	Fuping Movement		
Archaic	Hades	Fuping Stage	Land Nucleus Formation Stage			

Fig. 10.1.   List of phase and inheritance features of the tectonic systems in northwestern China.

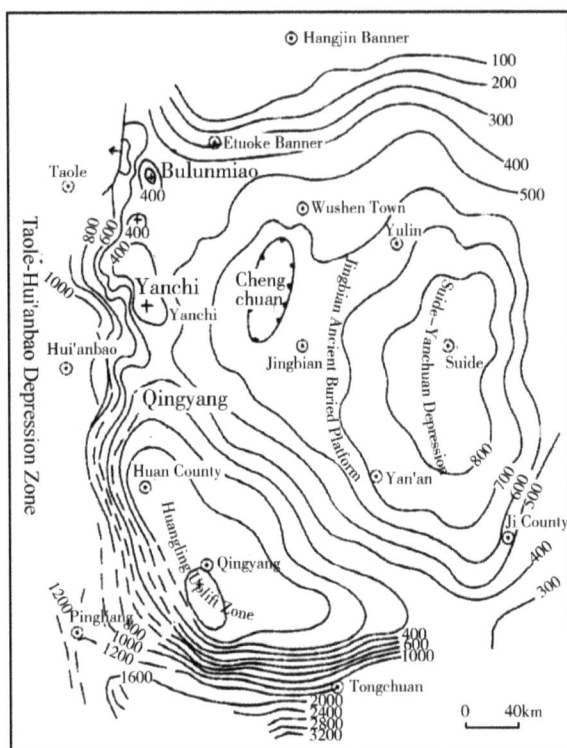

Fig. 10.2.   Paleotectonic map of the Ordos Basin in the Ordovician.

The deposition center of the Ordos Basin migrated to varying degrees in various periods from the Cambrian. For example, the deposit center of the basin was on the southwest in the Ordovician but migrated to the western and eastern regions (Figs. 10.2 and 10.3).

The Sichuan Basin still has tectonic migration features of different periods, which is manifested by the inconsistent deposition centers of different periods from the Sinian.[2] For example, the deposition center was located on the western margin of the basin in the Late Triassic but migrated to the northeast of the basin in the Jurassic (Figs. 10.4 and 10.5).

(4) Difference
The evolution of each tectonic system has different features and different activity intensities in different regions. For example, the Tarim Basin is composed of the Luntai faults of the E-W-trending system. During its

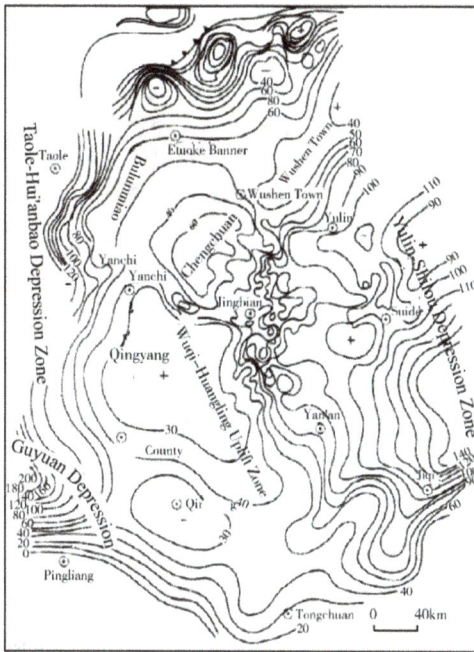

Fig. 10.3.   Paleotectonic map of the Ordos Basin in the Carboniferous.

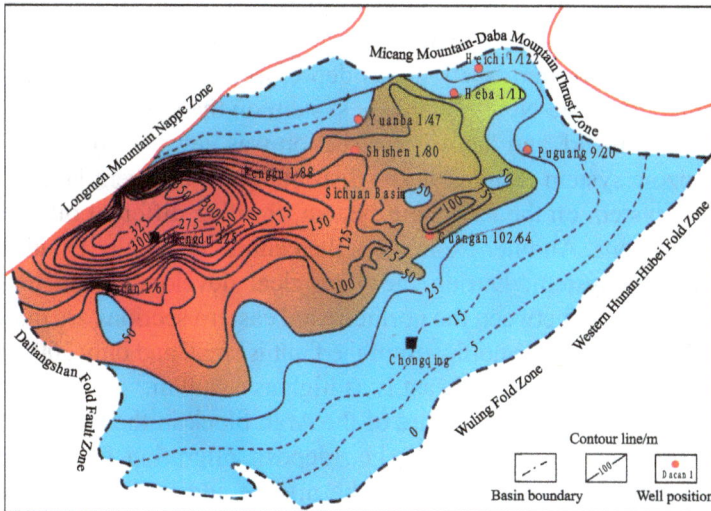

Fig. 10.4.   Thickness contour map of the dark mudstone of the 5th section of the Xujiahe Formation in the Sichuan Basin.

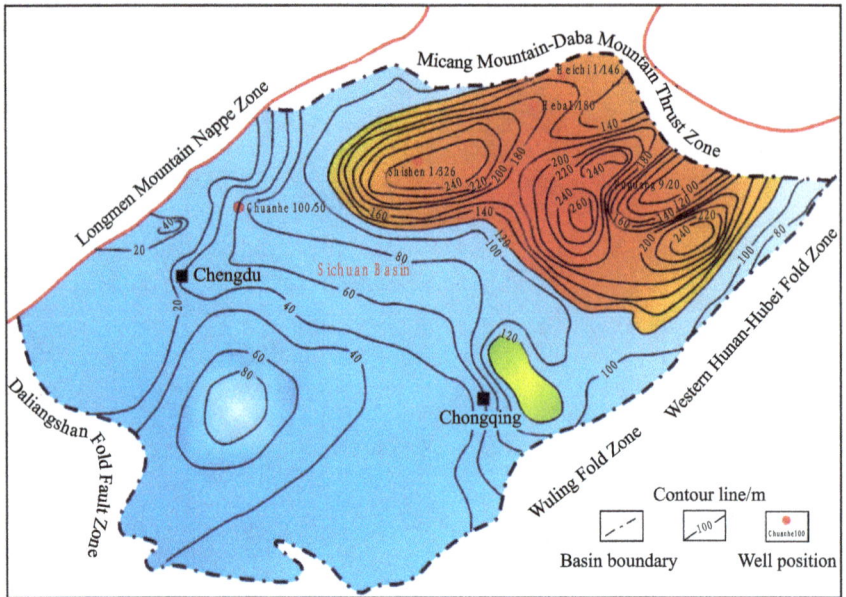

Fig. 10.5.   Thickness contour map of the lower Jurassic dark mudstone in the Sichuan Basin.

formation and evolution, it has experienced many activities, and the activity intensity and nature vary in each phase, showing different control effects on oil and gas accumulation. As the eastern section of the fault was uplifted and the western section was depressed, resulting in the distribution of the Paleozoic-New Metaphyte system on the upper layer, the eastern section was old, and the western section was new. The distribution of the Mesozoic system resulted in the distribution of different layers of the Mesozoic system on the Paleozoic system, forming different oil accumulation conditions. Another example is the central Tarim No. I fault in the N-W-trending system. During the geological development history, the activity intensities between the northern and eastern sections are quite different. The fault distance northwest of the fault is small and only 300–500 m, while the fault distance southeast is as high as 1,000 m.[3]

The deposit migration feature of the Arab block in the middle East is very obvious in various periods, i.e., deposit migration features in the early Silurian and early Devonian, the middle and late Eocene and the Oligocene (Figs. 10.6–10.9).

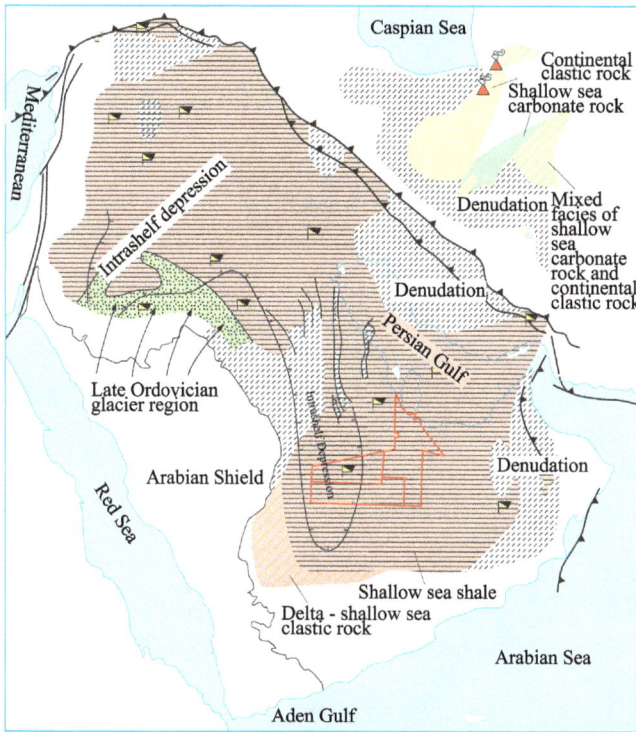

Fig. 10.6.   Deposit map of the Arabian block in the early Silurian.

In addition, in the Marakaipo Basin in South America, the migration feature of deposits was very obvious in the Late Cretaceous–Oligocene (Fig. 10.10).

(5) Conversion

The conversion occurring between compression and tension and between uplift and depression is another feature in the developmental history of the tectonic system. For example, the Shayazhan uplift in the E-W-trending tectonic zone north of the Tarim Basin was a depression zone in the early period of the early Paleozoic and converted into an uplift zone in the Silurian–Devonian. The Yanan fault on its northern margin was a south-dipping compression-torsional fault zone before the Himalayan but was converted into a south-dipping tensional fault zone since the Himalayan period.[4]

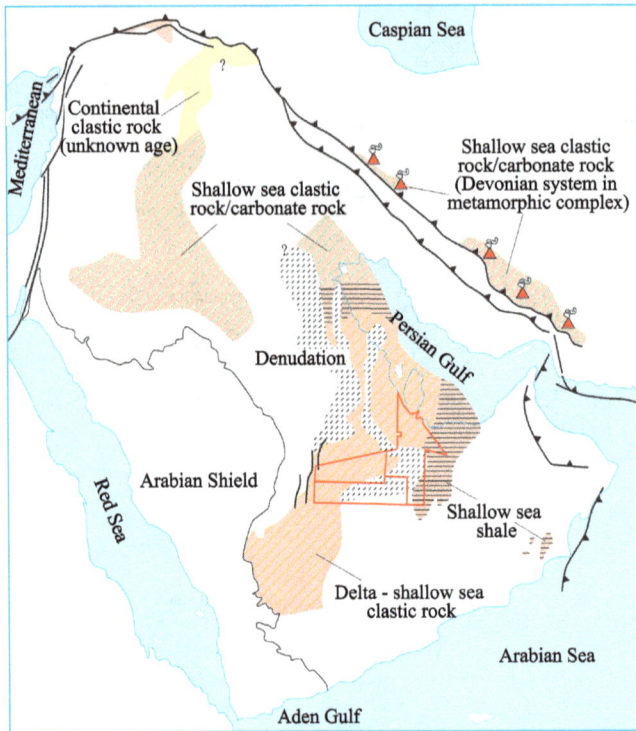

Fig. 10.7.   Deposit map of the Arabian block in the Early Devonian.

In addition, the Helan Mountain fault zone on the western margin of the Ordos Basin and the Longmen Mountain fault zone in the Sichuan Basin show obvious differences in the fault activity intensity of the southern-central and northern sections.

10.2 Compound of Tectonic Systems

Over the long geological period, the crust has undergone many tectonic changes. There are often several different types of tectonic systems in the same region. Therefore, the various tectonic phenomena seen in the field are the comprehensive result of tectonic deformation caused by previous tectonic movements. A series of tectonic traces or tectonic zones, as well as certain shapes of rock blocks or land blocks, can be formed in a tectonic

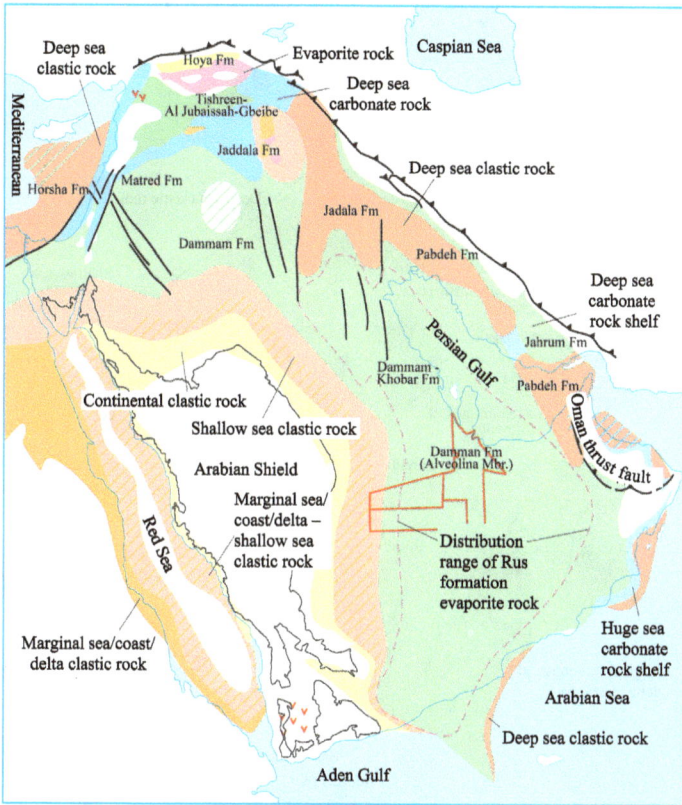

Fig. 10.8.   Deposit map of the Arab block in the middle and late Eocene.

movement that develops in a certain way. They are distributed in a certain pattern, which is restricted and accommodated by existing tectonic zones or blocks and utilized or reformed in later tectonic movements. Some new tectonic traces or tectonic zones may be formed in land blocks or rock blocks with weak tectonic traces from the previous time period,[5] while the latter can become a component of the latter in some regions, and similar situations may occur in tectonic traces with different sequences formed in the same tectonic movement. The above is a tectonic compound phenomenon.

(1) Miter Compound
If two tectonic zones are very similar in direction, it will be difficult to distinguish them in some sections. However, if they are traced along

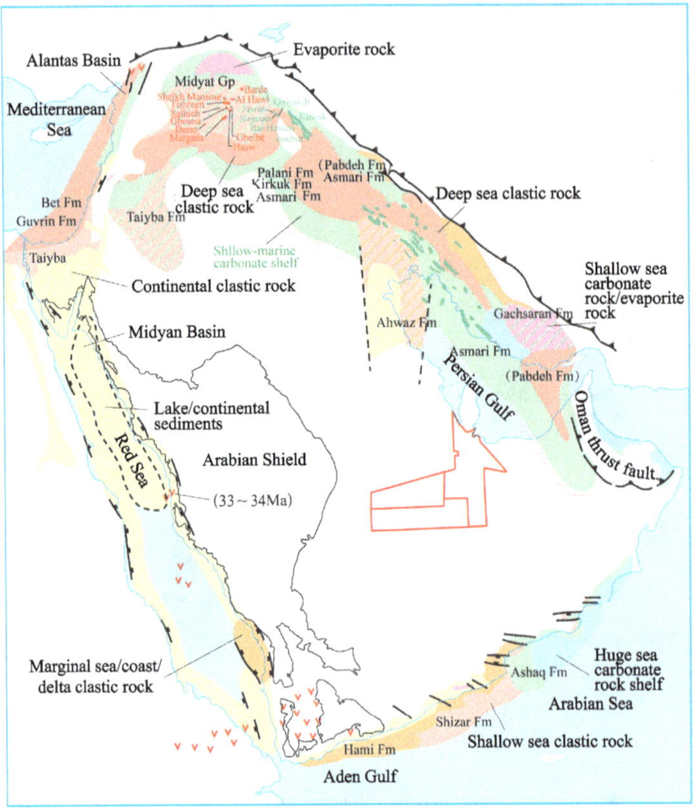

Fig. 10.9.   Deposit map of the Arab block in the Oligocene.

the direction, they are gradually separated, forming what is called a miter.

In Tarim and its surrounding regions, the compound relationship between the N-W-trending and E-W-trending systems belongs to this type. In the Tian Mountain region, the Tian Mountain E-W-trending system is mitered with the Borohoro E-W-trending zone. The Bolokonu zone is an extremely strong compression and rotation-torsional tectonic zone. Due to their compounding, it is extremely obvious that a compression and compression-torsional tectonic system with Zhongtian Mountain as the axis is formed in the region.

On the northeastern margin of the Tarim Basin, the N-W-trending Kongque River slope may be a component of the primary uplift zone of

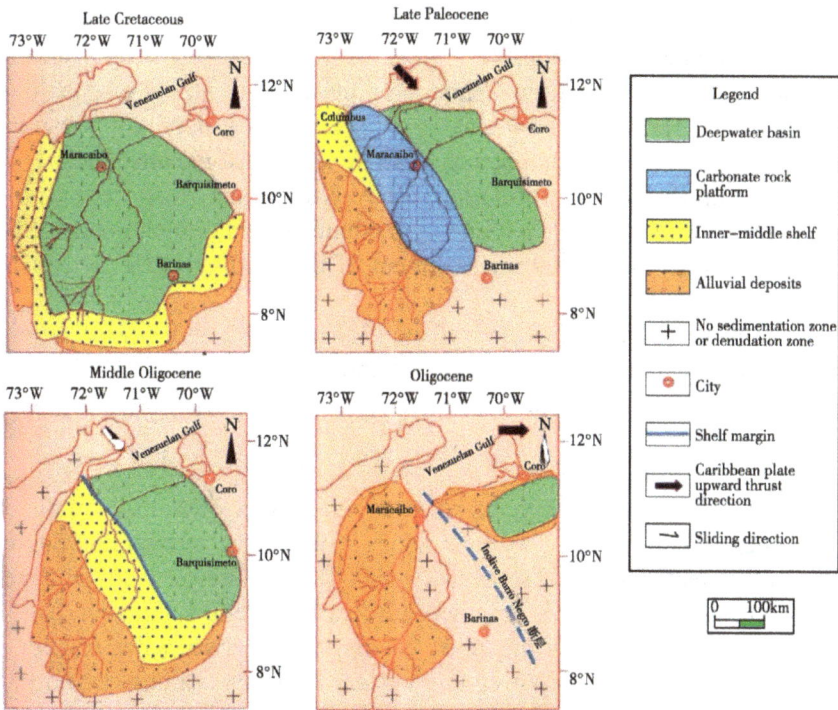

Fig. 10.10.   Paleogeographical distribution of deposits in the Marakaipo Basin of South America in the Late Cretaceous–Oligocene.

the N-W-trending system and is compounded with the South Tian Mountain E-W-trending zone, forming the northern slope of the depression region northeast of the Tarim Basin (Manjiaer Depression). The Bachu–Katake zone is uplifted in the center of the Tarim Basin and strengthened further by the compound of the E-W-trending uplift zone and the slope of the N-W-trending system, resulting in obvious presentations of N-W-trending compression-torsional faults in the region, i.e., the Tumshock fault, Gudong Mountain fault, Mazatage fault, Hemiros fault, Karashay fault, central Tarim No. 1 fault, central fault barrier zone, central Tarim No. 10 fault, etc.[6] With the further development of the N-W-trending system, the Shuntuogole N-W-trending low uplift between the Awati depression and the Manjiaer depression and the Asha No. 4 fault and Aman No. 2 fault on its side are also formed, which are its low-level tectonic components. Because of the emergence of the N-W-trending system, the Aman depression zone is divided into the Awati depression and the Manjiaer depression.

Based on the distribution pattern of the N-W-trending system, the southwestern Tarim depression is also regarded as a depression region (Paleozoic) with E-W-trending and N-W-trending tectonic attributes. The Gonggeer uplift (central uplift) in western Kunlun may also be the primary uplift zone of the N-W-trending tectonic system, which is now difficult to identify due to the involvement of the Pamir reverse S-shaped tectonic system.

Due to the miter compound of the N-W-trending and E-W-trending systems in the Tarim block, the E-W-trending blocks are superimposed with the N-W-trending blocks, controlling the Tarim basin together.

(2) Reverse Compound

The so-called reverse compound refers to a relatively obvious compound phenomenon between two tectonic zones with a large intersection angle (generally greater than 45° or even up to 90°). If the newer tectonic zone is a fault, the older tectonic zone will be clearly cut and staggered.

A typical example is the reverse compound between the rotation-torsional faults of the N-W-trending system distributed along the northwestern margin of the Tarim Basin and the southern Tian Mountain arc distributed along the northern margin of the basin, especially for the following sections.

(a) *The peculiar Kuzgunsu depression trough*: The S-W-trending southwestern Tian Mountain fold zone is suddenly intersected by the N-N-W-trending Kangsu–Miya fault, forming a deep trough. It extends in the S-S-E-trending direction and is connected with the Kuzigongsu fault and then is inserted east of the West Kunlun Mountains to the south.[7] In this trough, the Jurassic system is well preserved, with a thickness of more than 4,000 m, and the hydrocarbon source rocks are well sealed. This example is a relatively obvious reverse compound phenomenon.

(b) *N-N-W-trending Keping arc fault nappe zone of the N-W-trending system*: The nappe collision zone (the Keping–Shajingzi fault zone) between the Keping arc and the Bachu uplift northwest of the Tarim Basin is N-N-E-trending and is reversely intersected by the N-N-W compression and rotation-torsional fault of the N-W-trending system. Among them, the main faults of the N-W-trending system include the Pulv–Selibuya fault zone, the Pichaksun fault and the Acha fault zone, followed by the Tonggang fault, the Qiaoniergai fault, the Bachu fault and the Sanchakou fault. Because of their development, not only is the Keping arc intersected but the northwestern section of the Bachu uplift is also transformed and distorted. The above compound phenomenon occurred in the late Mesozoic–Cenozoic.

(c) *Kala Yuergun fault zone*: This fault zone is located at the intersection of the Mashi depression and Baicheng depression. Due to the intersection of the N-W-trending rotation-torsional fault, the N-E-E-trending tectonic zone is disturbed in this region and is compounded with the derived branch tectonic zone into a herringbone-shaped tectonic zone during the southward extension process of the Kala Yuergun fault.[8]

In addition, in the Korla region on the northeastern margin of the basin, the NSW-SEE-trending ancient fold zone (South Tian Mountain arc) is intersected by the N-N-W-trending Korla rotation-torsional fault in reverse compound and miter compound forms. At the same time, the intersected ancient tectonic zone is twisted clockwise. It is also a compound phenomenon involving the N-N-W-trending tectonic system.

The abovementioned reverse compound phenomenon is a local phenomenon occurring on the western and northern margins of the Tarim Basin in the late Mesozoic–Cenozoic, resulting in some tectonic interference with the abovementioned regions.

(3) Overlapping Compound
This compound phenomenon occurs mostly in regions with large-scale uplifts or depressions. Some or all sections of a complete tectonic system can be strengthened by large-scale uplift or weakened by large-scale depression. This kind of strengthening and weakening does not refer to the strength of the original tectonic zone but is a superficial phenomenon reflected by the overlapping compound effect.[9]

The Tarim platform was folded and finalized in the late Hercynian period. In the Mesozoic and Cenozoic tectonic development stages, due to the formation and rapid depression of the Kuqa piedmont depression zone, the Tarim platform marginal uplift (Shaya paleo-uplift), the north Tarim platform marginal depression (marginal sea basin), the Aman paleo-depression region and the central uplift zone of the original E-W-trending system were depressed and buried, but they were not weakened due to depression. The current seismic exploration results have confirmed that the above tectonic components of the E-W-trending system still exist. This phenomenon is the overlapping compound phenomenon of the southern Tian Mountain arc and the E-W-trending system.

(4) Transformation Compound
Based on the predictability of the distribution of tectonic systems, it is believed that some E-W-trending tectonic components are also developed south of the Tarim platform, i.e., the South Tarim depression zone and the South Tanan marginal uplift zone. Now, only one incomplete Tangubas depression is visible, and it is difficult to identify other E-W-trending zones. Visible Mesozoic and Cenozoic tectonic traces show a large reverse S-shaped depression zone southwest of the basin, namely, the Pamir S-shaped marginal depression zone, advancing from the piedmont zone to the northeast, and passing through the Maigaiti slope, after which it and the Bachu–Katake uplift are covered by the thick Mesozoic and Cenozoic system.[10]

References

1. Wang Jiashu. *Timan-Baichaola Oil and Gas Bearing Basins*. Beijing: Petroleum Industry Press, 1991.
2. Wang Jun, Wang Dongpo, C. A. Ushakov *et al. Formation and Evolution of Sedimentary Basins in Northeast Asia and Their Petroleum Prospects*. Beijing: Geological Publishing House, 1997.
3. Wang Zhixin, Jin Zhijun. *Petroleum Geological Characteristics of the Siberian Platforms and Its Marginal Depressions*. Beijing: China Petrochemical Press, 2007.
4. Xiong Liping. Tectonic Evolution in the West Africa and its Control on Hydrocarbon Accumulation. *Oil & Gas Geology*, 2005, 25(6): 641–646.
5. Bai Guoping. *Petroleum Geological Characteristics of Middle East Oil and Gas Regions*. Beijing: China Petrochemical Press, 2007.
6. Chen Tingyu, Shen Yanbin, Yue Zhao *et al. Geological Development of Antarctica and Evolution of Gondwana*. Beijing: Commercial Press, 2008.
7. Deng Xiguang, Zheng Xiangshen, Liu Xiaohan. Discovery of Gravel-bearing Mudstone Layers in Livingston Island, Antarctica and Its Geological Significance. *Polar Research*, 1999, 11(3): 169–178.
8. Zhang Kang, Zhou Zongying, Zhou Qingfan. *Petroleum and Natural Gas Development Strategy of China*. Beijing: Petroleum Industry Press, 2002.

9. Basu, D. N., Bancrjec, A., Tainhanc, D. M. Source Areas and Migration Trends of Oil and Gas in Bombay Offshore Basin, India. *AAPG Bulletin*, 1980, 64(2): 209–220.
10. Beiriein, E. P., Arne, D. C., Keay, S. M. *et al.* Timing Relationships between Felsic Magmatism and Mineralization in the Central Victorian Gold Province, Southeast Australia. *Australian Journal of Earth Sciences*, 2001, 48: 883–899.

Epilogue

Yuzhu Kang and Shuwen Xing

(1) The key driving forces of the formation of tectonic systems include changes in the Earth's rotation speed, the influence of celestial masses on the Earth, radioactive elements in the crust, the heterogeneity of crust thickness, and differences in crust density. The combined effects of these factors have created multidirectional stresses in different regions in different periods, resulting in the formation of different tectonic systems.

(2) For the first time, eight major tectonic system types in the world were systematically established as follows: (i) the E-W-trending tectonic system; (ii) the N-S-trending tectonic system; (iii) the N-E-E-trending tectonic system; (iv) the N-N-E-trending tectonic system; (v) the N-W-trending tectonic system; (vi) the epsilon-shaped tectonic system; (vii) the S-shaped or reverse S-shaped tectonic system; and (viii) the rotation-torsional tectonic system. The main tectonic systems are E-W-trending and N-S-trending.

(3) The evolutionary features of each tectonic system were classified as follows: phase, inheritance, difference, migration and conversion, indicating the complexity of each tectonic system.

(4) The formation and evolution of various large and small blocks are controlled by tectonic systems, and the formation and evolution of the tectonic systems are controlled and affected by blocks. Current global

tectonic patterns and changes and evolution in the land and sea are created by these interactions.

(5) The formation of deposits and prototype basins as well as the formation, transformation and finalization of global energy minerals and metal minerals were controlled by the formation and evolution of tectonic systems in various periods. The distribution of minerals was controlled by tectonic systems in a very regular manner. The regional distribution of minerals is controlled by primary and secondary tectonic systems, and the distribution of mineral deposits and oil and gas fields is controlled by the third and fourth levels of tectonic systems.

Index